Environmental Policy and Governance in China

Hideki Kitagawa
Editor

Environmental Policy and Governance in China

Springer

Editor
Hideki Kitagawa
Faculty of Policy Science
Ryukoku University
Fushimi-ku, Kyoto
Japan

ISBN 978-4-431-56488-1   ISBN 978-4-431-56490-4 (eBook)
DOI 10.1007/978-4-431-56490-4

Library of Congress Control Number: 2016962720

Printed on acid-free paper

This Springer imprint is published by Springer Nature
The registered company is Springer Japan KK
The registered company address is: Chiyoda First Bldg. East, 3-8-1 Nishi-Kanda, Chiyoda-ku, Tokyo 101-0065, Japan

Preface

Human-induced environmental pollution and destruction as a result of rapid economic growth has become an issue of serious concern affecting people's health in China. Although some improvements have been made since the Communist Party and the Central Government established and enacted various laws and policies, the present situation remains unpredictable and daunting. In the background of this paradoxical situation, the significant tendency toward "rule by men" in various environmental concerns still firmly remains, giving the impression that China is a nation of "rule by law," not a nation of "rule of law."[1]

In the past, many obstacles against law and policy administration have been noted. These include local protectionism practiced by the local governments to protect polluter firms and corporations, regulatory issues such as lack of information disclosure, limited the amount of authority of the Ministry of Environmental Protection, penalties for activities that violate the ethical code or environmental law, and non-independence of judicial power from the administration. In my recent field investigations, I found these issues remained unchanged. At the same time, the general lack of environmental awareness among the Chinese citizens is also a major stumbling block to environmental improvement. In view of this, environmental education and the dissemination of environmental knowledge are essential. In short, circumventing the current environmental quagmire calls for a mix of policy instruments, as it would have been impossible to tackle all the problems and issues as mentioned with just one single instrument.

With this view in mind, this book attempts to examine how stakeholders such as the legislators, administrators, corporations, citizens, and environmental NGOs cooperate to solve the difficult environmental issues confronting China from multiple perspectives. The main focus of examination is mainly related to the

[1] China is not a country characterized by the "rule of law," a concept developed in England on the premise of liberty and democracy. By contrast, China has a tendency to use rules as a tool for administration, better described as "rule by law."

development and enforcement of environmental law and policy in China. Some of the major issues covered environmental protection law, soil pollution, the target responsibility system in pollution prevention policies, environmental resource tax, multilayered pollution problems, visits and letters in association with environmental management, public participation in EIA, and environmental NGOs, among others. In the first half, we review how these issues are approached from the aspect of government systems and policies, and in the latter half, we discuss approaches from the aspect of citizens.

Chapter 1 introduces eco-civilization, Xi Jinping Government's environmental policies from the decision on the Third Plenary Session of the 18th Central Committee of the Communist Party of China and the amendment to the Environmental Protection Law enacted in January 2015. This chapter also examines issues for the improvement of environmental governance. In Chap. 2, the author, who participated in the amendment process of the Environmental Protection Law, introduces the detailed discussion on the constitutional process of the amendment law. More specifically, this chapter reveals that the principle of transition from the priority on the economy to environmental protection does not necessarily go hand in hand with environmental administrative transformation.

Chapter 3 discusses soil contamination, an issue that, though previously suspected, has recently been made clear through actual environmental studies. Regulations related to soil contamination have been partly enacted, but they were not sufficient to solve the food safety issues, which affect the health of the citizens. In acknowledging the need for the establishment of effective mitigating schemes, this chapter discusses what constitutes an effective soil protection law system. Some of the issues discussed include mechanism for administrative improvement, an analysis of responsibility system, and financial support for environmental improvement. Because of rapid economic development in recent years, China has become a leading consuming country of fossil fuels. Currently, it is the world's largest carbon dioxide emitting country. It also confronts with acute PM2.5 air pollution problem. Chapter 4 discusses the background of the enactment of the target responsibility system (TRS) for environmental protection to overcome these issues, the process of allocating targets to local governments, its legal and political position, and other related issues.

Chapter 5 takes up the coal resource tax issue, which serves as an effective means in reducing the consumption of fossil fuels as discussed in the previous chapter. This chapter also analyzes the impacts of the tax system on fiscal revenue, resource conservation, and energy savings for the period of 1994–2011. The goals and challenges of the 2011 coal resource tax reform, which changed the tax system from a volume-based system to a fixed-rate ad valorem system based on sales, are also the focus of this chapter. Chapter 6 focuses on the discussion of the limited relief available to pollution victims in environmental litigation. It clarifies the fact that decisions regarding an environmental dispute made by the local committee of the CPC could have substantial impact on the decisions of the local court, while the EPA of the upper level local government or the upper level local court could be disregarded by the local court.

Chapter 7 discusses the effectiveness of China's traditional environmental complaint letters and visits system. In the current situation with courts taking cognizance of three percent of all environmental litigations in China, this chapter examines the importance of the administration body that handles environmental complaints and alternative dispute resolution (ADR). The establishment of an organization with authority to work out solutions to environmental disputes is suggested.

Chapter 8 introduces public participation in the Environmental Impact Assessment (EIA). It is pointed out that the EIA system introduced in the late 1970s did not include public participation. Although public participation in EIA was introduced in the late 1990s, it was extremely vague and hence defeating its very purpose of participatory improvement. In 2006, the Interim Measures on Public Participation in EIA were enacted. However, issues such as the lack of effectiveness and the absence of trust in public participation in recent years constitute new obstacles to participatory enhancement. China's initial legislative amendment shows an intention to improve important issues such as the scope of legal rules and implementation, and greater consideration for public opinion in environmental issues. Chapter 9 covers the case of chromium pollution by vitamin K manufacturers in Yunnan province. More specifically, it exposes the actual situation of multitiered pollution of the industry concerned. This chapter also touches on the role of NGOs in helping to solve this environmental pollution issue based on actual case studies in Yunan and Hunan provinces.

The last chapter, Chap. 10, discusses the history and the roles of environmental NGOs in China. Twenty years has passed since grassroots NGOs were born in China. They have dedicated themselves to solving environmental problems through the development of NGO networks, environmental prosecution and the enhancement of cooperation and communication with pollution victims and supporters in Japan.

Overall, the book provides extensive analysis on various important environmental issues confronting China and suggests mitigating measures to contain the problems concerned. We hope this book will prove useful to not only scholars who study environmental governance in China but also readers who are interested in environmental issues in China.

Faculty of Policy science
at Ryukoku university
Kyoto, Japan

Hideki Kitagawa

References

Kitagawa, Hideki "Environmental Law and Policy and its Implementation of China" (from "Environmental Law and Policy and Governance of China" Introduction) Koyoshobo, January 2012.

Kitagawa, Hideki "Effects and Issues of Chinese Environmental Law's Implementation for 30 years-Mainly on the Amendment to the Environmental Protection Law and the Environmental Dispute Institution" (from chapter 12 of Hideki Kitagawa et al. edition "Development and Change of Modern Chinese Law", Seibundou) July 2013.

Wang, Jing, Zhang, An, Yan, Houfu, "Environmental Law Administration is Difficult. Where Difficulty lies ?-A report of Questionnaire to staffs of Environmental Protection Department in China", Environmental Law in China, Law Press China, 2009.

Contents

Contributors

Akihiro Chiashi is researcher at the Kyoto University Asian Studies Unit, Japan. His research interests are in environmental politics and economics, environment and development, environmental governance, and policies in China and Asian countries. His current research focuses on the political economy and governance of the environmental pollution in China.

Hideki Kitagawa is professor in the Faculty of Policy Science and head of the Research Institute for Social Sciences, Ryukoku University, Kyoto, Japan, researches environmental law and policy of Japan and China. His current research focuses on Chinese environmental governance, particularly how to solve difficult environmental issues in cooperation with corporations, NGOs, citizens, etc.

Jiro Sakurai is associate professor of Chinese studies at Kobe City University of Foreign Studies. He wrote mainly about Chinese environmental law and policy, mostly focusing on litigations as one of some measures to relieve the victims of environmental damages. He wrote only in Japanese, so the main challenge for him to tackle with is to write papers in English.

Jin Wang is professor at Peking University Law School and the director of Resources, Energy and Environmental Law Institute, Beijing, China. His current research focuses on environmental law and policy and nuclear policy and law.

Shekun Wang is associate professor at Peking University Law School, as well as research fellow of Resources, Energy, and Environmental Law Institute (REELI) at the same university. His current research focuses on the enforcement of environmental protection law in China and the allocation of the rights to natural resources.

Shijun Zhang is professor at the Law School of Shandong University and the director of the Research Institute of Economic Law. He is standing director of China's Society of Environment and Resources Law. His current research focuses on environmental lawsuit, environmental dispute resolution, climate change, environmental governance, etc.

Xiaobo Zhao is lecturer at the School of Law and Justice, University of Southern Queensland, Australia. His research interest focuses on Chinese environmental law, comparative contaminated land law, and torts. His monograph titled "Developing an Appropriate Contaminated Land Regime in China: Lessons Learned from the US and UK" was published by Springer in 2013. He is also the author and coauthor of a couple of peer-reviewed journal articles and book chapters.

Yanmin He is a postdoctoral fellow of the Institute of Economic Research at Kyoto University, Japan. She completed her PhD at Kyoto University (2014) with a study of China's resources taxation. Her current research interests include Chinese economy and Chinese environmental governance and focus especially on the energy and environmental fiscal reform for a sustainable society.

Yasushi Aikawa is associate professor in the Faculty of Business Administration at Tottori University of Environmental Studies, Tottori, Japan. His research interests are in the relationship between environmental pollution and society in China and Japan. His current research focuses on the way of Chinese environmental NGOs solving environmental pollution and their cooperation with Japanese environmental NGOs.

Zhen Jin is a fellow with the China Research and Communication Center (CRCC) at Japan Science and Technology Agency (JST). His research interests cover various aspects of low-carbon city policies, especially energy-saving policies and air pollution control policies in China. His areas of specialization are administrative law and environmental law, and he obtained his doctorate from the Law Department of Kyoto University in March 2009.

Chapter 1
Environmental Policy Under President Xi Jinping Leadership: The Changing Environmental Norms

Hideki Kitagawa

Abstract Across China, exacerbating air quality deterioration, water pollution, and land degradation are serious problems that often lead to episodes of violent protest and public unrest. It goes without saying that the prevention and control of these crisis-ridden environmental problems pose one of the most daunting tasks confronting China today. Against this backdrop, the purpose of this chapter is to examine the characteristics and effectiveness of China's recent environmental policies in addressing its mounting environmental problems. These include China's new blueprint for the promotion of eco-civilization, the Chinese Communist Party's environmental resolutions, and environmental law amendments. The chapter concludes by suggesting ways to addressing the current environmental woes facing China today.

Keywords China's environmental policy • Environmental governance • Eco-civilization • Environmental protection law amendment • Execution of law

Abbreviation

CPC Communist Party of China

1.1 Introduction

For the past few decades, China has been achieving an impressive economic growth at an annual average of more than 9 % since it implemented its market reform in 1978. As the economy is growing rapidly, various environmental issues are spreading across the nation: air pollution such as PM2.5, mainly in Beijing and other northern regions; water pollution originating in industrial wastewater from factories

H. Kitagawa (✉)
Faculty of Policy Science, Ryukoku university, Kyoto, Japan
e-mail: kitagawa@policy.ryukoku.ac.jp

H. Kitagawa (ed.), *Environmental Policy and Governance in China*,
DOI 10.1007/978-4-431-56490-4_1

which led to the proliferation of "Cancer Village" across China; food safety crisis caused by soil contamination and ground water pollution, and environmental issues related to landfill wastes. Given the dire environmental situation in China, the country is sometimes called as "a department store of environmental issues" (Koyanagi 2010). These issues have been spreading nationwide and emerging with a multilayered structure; therefore, they are extremely difficult to solve.

This paper analyzes the characteristics and issues of recent Chinese environmental policies of the Xi Jinping government, focusing on the development of eco-civilization; "the resolution on the reform of several serious issues" in the third conference of Communist Party's central comittee held in November of 2013 and the amendment to environmental protection law.

1.1.1 Proposal of Eco-civilization

Xi Jinping took the reins as both a general secretary and a president of the central military committee in the 18th general meeting and the following 1st Central Committee Conference of Chinese Communist Party (hereinafter referred as CCP), which means that Xi is an actual leader with political power.

In this meeting, the president of China, Hu Jintao (at that time), suggested an ideological expansion from socialism with the Chinese characters "四位一体"(four in one ideology)—economy construction, politics construction, culture construction, and society construction—to "五位一体"(five in one ideology), adding eco-civilization construction into it and spreading it across all other ideological fields and processes. This ideological integration constitutes the idea of respecting and protecting nature as appeared in Hu's report adopted at the 18th Party Congress.[1]

Chapter 8 of this report, "Making Great Efforts to Promote Ecological Progress," declares the following basic principles of environmental protection : ① committed to constructing an aesthetically pleasing China, ② adherence to the fundamental national policy on resource saving and environmental protection, and ③ devotion to high-priority advocacy of priority to conserving resources, environmental protection, and natural restoration.

Moreover, it places emphasis on the following points: ① to enhance spatial development across China and to control the speed of economic growth and to regulate spatial development structure in the country, ② to promote an all-encompassing approach to resource saving and conservation, ③ to reinforce ecosystem and environmental protection efforts, and ④ to enhance the creation of ecological civilization system which underpins environmental protection efforts.

[1]"In building socialism with Chinese characteristics, we will surely complete the building of a moderately prosperous society in all respects," November 8, 2012.

What does the "eco-civilization" that is emphasized here mean? This idea was proposed by the Hu Jintao government in 2006 and was defined on the Internet as the next stage in the development of human civilization that comes after industrial civilization. It also means the fruits of the material and spirit that mankind obtains through a rule of harmonized development among man, nature, and society. The Chinese traditional idea of 天人合一思想 (harmony between man and nature)[2] is thought to be an important cultural origin of this idea. However, the meaning of this idea has not been clearly defined, and eminent people from various fields are discussing it. It seems that it is taking a significant amount of time to define it.

President Xi Jinping announced the Eight-Point Regulation at a meeting of the Politburo of the Communist Party of China in December 2012 which aimed to enhance work ethics and to make the party closer to the Chinese community at large.

During the two conferences in March of 2013, Prime Minister Li Keqiang announced the prohibition to construct new government buildings, to reduce work-force size, to absolve from receiving guests or traveling abroad at public expense, and to decrease the number of public cars during the term of the present government.

The Xi Jinping government took office in November 2012, and it can be said that his political ideology was ahead of his actual actions and that it was too premature to expect an immediate reform of concrete environmental policy. His unique policies are expected to be enacted in his next term. However, in an interview conducted in May 2013 with an officer, in charge of environmental policy for the Environmental Protection Bureau, this officer who wished to remain anonymous, informed me what we could be expected based on the two statements announced by the leaders. One of them is Xi Jinping's statement delivered at a press conference in relation to the member of the Standing Committee of the Central Political Bureau on November 15, 2012, "people expect a good environment, which is our goal," and the other is Li Keqiang's statement that "Chinese development has not been restricted owing to the environment in its 3000-year history. Environmental issues are prominent ones that attract the serious attention of our government." Moreover, at a workshop conducted in relation to ecological civilization of the Central Political Bureau on May 24, 2013, Xi Jinping disclosed that environmental protection is an unyielding principle, and if someone brings about a serious outcome through policy without consideration of the environment, we will seek his/her liability throughout his/her life. People in general have a good impression of the government for this reason.[3]

---

[2]See 互动百科: Baike(http://www.baike.com/wiki/%E7%94%9F%E6%80%81%E6%96%87%E6%98%8E,2014.4.2).

[3]Interview was conducted with Mr. A from the Ministry of Environmental Protection on May 27, 2013. Beijing.

1.2 The Decision of the Third Plenary Session of the 18th Central Committee of CPC

At the Third Plenary Session of the 18th Central Committee of CPC, held in November 2013, the document, "Decision of the Central Committee of the Communist Party of China on Some Major Issues Concerning Comprehensively Deepening the Reform" was adopted.

The decision was met with applause by some intellectuals as groundbreaking, claiming that it contains important content concerning solutions for environmental issues and governance. For example, Chap. 14 of the document explicitly expressed its aim "to promote ecological progress" and to confirm its goal under the slogan of "ecological progress."

This chapter also provides regulations for the establishment of complete and integrated institutions and systems; implements the strictest source protection system, damage compensation system, and accountability system. This chapter also seeks to improve environmental treatment and ecological restoration systems; and uses set rules to protect the ecological environment. It also emphasizes the following four fields.

1.2.1 Improving the System of Natural Resource Property Rights and the System of Natural Resource Utilization Control

The party confirms the right to resources and unified registration for natural ecological spaces such as rivers, forests, mountain ranges, grasslands, undeveloped land, tidal land, etc. It establishes a natural resource property rights system with the property rights, rights and responsibilities defined clearly. It also establishes a spatial planning system and restricts boundaries for production and living and ecological space utilization. Moreover, it implements a system for conservation of energy, water, and land resources and for their intensive use and unifies the duties of the people who act as the owners of public natural resources.

1.2.2 Delimiting "the Red Line for Ecological Protection(红线)[4]"

The party unwaveringly implements a system of ecological functional areas through the establishment of a system of land and space development and

[4]This means a settled principle on which they cannot compromise.

protection. It also promotes development strictly in accordance with the ecological functional definition of this new environmental protection ideology and establishes the national park system. The party also establishes a mechanism to monitor and provide early warning to the risks of ecological perturbation of the carrying capacities of resources and the environment. It also implements restrictive measures for regions where water and land resources, the environment, and oceanic resources have been excessively exploited.

It also cancels regional GDP assessment for regions under restrictive ecological development and for national key counties which are ecologically fragile. To reinforce environmental conservation efforts, the party also reforms policies that emphasize on economic growth.

The party also explores ways to compile a natural resource balance sheet, implements natural resource audits when leading officers left their positions, and establishes a lifelong accountability system for ecological and environmental damage.

1.2.3 Implementing Compensation Systems for Use of Resources and for Damage to the Ecological Environment

The party accelerates pricing reform for natural resources and their products to give full reflection of their market supply and demand trends, the extent of resource scarcity, ecological and environmental damage costs, and restoration benefits. In relation to the utilization of all natural and ecological resources, it adheres to the principle that whoever uses the resources must pay for them, and that whoever pollutes the environment and damages the ecology must pay compensation. It also gradually imposes resource taxes on the possession and utilization of all natural and ecological resources.

The party also expands the green areas in places where farmland is converted back to forests and grassland, adjusts the utilization of farmland in heavily polluted areas or areas where underground water has been overexploited, and implements an orderly rehabilitating system for farmland, rivers, and lakes. It also establishes effective comparative pricing mechanism to regulate industrial land and residential land prices and raise the price of land for industrial use. In persistently adhering to the principle that he who profits should compensate, it has been able to improve the ecological compensation mechanism for key ecological functional areas and promotes the establishment of a horizontal compensation system between regions and the development of the environmental protection market. The party also implements a trading system for energy conservation, carbon emission, and waste discharge and water utilization rights. In addition, it further establishes a market-oriented mechanism to attract private capital to ecological and environmental protection and implements third-party treatment of environmental pollution.

1.2.4 Reforming the Ecological Protection Management System

The party establishes an environmental protection system that strictly supervises the emission of all pollutants, a joint interregional mechanism for comprehensive land and marine ecosystem protection, and restoration and pollution prevention. It will improve the operation and management system for the state-owned forest regions and press ahead with the reform of collective forest rights. It publicizes environmental information in a timely manner, improves reporting system, and strengthens social supervision. In addition it will improve the licensing system for pollutant emission and implement quantity control system for pollution emission by enterprises and public institutions. The strict implementation of the compensation system for those responsible for ecological or environmental damage and investigation of any criminal responsibilities in accordance with the law are added.

Apart from the above, the adopted document as mentioned above also regulates some notable issues in relation to ecological governance. For example, in Chap. 4, "Accelerating the Transformation of Government Functions," prescribes reforms for an investment system, reinforces market access standards concerning energy, land and water conservation, the environment, technology and security, etc. Also, a long-term mechanism is scheduled to be established and improved for preventing and reducing excess production capacity. The party also aims to improve the development progress evaluation system and to rectify the bias on political achievement evaluation processes which are fundamentally based on economic factors. It also reinforces other evaluation indicators, such as resource consumption, environmental damage, ecological benefits, excess production capacity, technology innovation, production safety, and new debts, while more emphasis is put on improving standards for energy saving standards and environmental protection.

In addition, the adopted document also regulates the notable reform of the judicial administration system. This is explicitly stated in Chap. 9, "Promoting the Rule of Law." The party will unify the management of staff, funds, and properties of the courts and procuratorates below the provincial level and explores ways to establish a judicial jurisdiction system that is appropriately separated from the administrative divisions to ensure that the state laws are enforced properly and uniformly. It has been pointed out that there is a tendency toward local protectionism (Wang 2011, page 309).[5] It may well be that in all over the state where the finances of the local courts are paid by the local government, and the personnel of the local court are appointed by local people's congress; the judgment delivered is expected to be influenced by economic factors, that is, local profit margin levels. We can anticipate this reform will take a step toward enhancing the independence of the judge, once the superior courts manage the lower courts directly.

[5]The tendency is that the leaders in local government try to protect the economic benefits of the area.

1.3 Amendment to the Environmental Protection Law

The 11 provisions, such as the amendments of the land management law, environmental protection law, air pollution prevention law, and law on the prevention and control of water pollution, and the enactment of the Soil Pollution Control Act and the Nuclear Safety Act in the 68 provisions, clarified by the 5-year legislation plan (October 2013) of the Standing Committee of the 12th National People's Congress, are associated with "eco-civilization as mentioned above."[6] It is noteworthy that the amendment to environmental protection law (enacted in 1989), which is the basic law concerning environmental protection constitutes an effective legal instrument which ordains fundamental environmental policy. It also has the force in imposing heavy environmental responsibility.

The amendment to the environmental protection law had been enthusiastically discussed in the 8th National People's Congress (1993), where the Environment Protection and Resources Conservation Committee was established. In the following Standing Committee of the 9th–11th National People's Congress, the amendment had been included in the legislation plans, and research and investigation and administrative activities of law have been conducted dozens of times. Legal professionals and officials have been sent overseas to investigate tendencies of legislation in the ecological field and implemented programs for the development of human resources. Professor Jin Wang from Peking University pointed out that the difficulties with this issue can be seen from the fact that the environmental law could not be revised in spite of these efforts (Wang et al. 2012: page 51).

Again, at the Standing Committee of the 12th National People's Congress, the process to amend the legislation had already started with the first revision draft for proposal of the environmental protection law published in August 2012, and the hearing of opinion held. The outline of the draft was: "divisions and administrations in the field of oceanic, fishery, transportation, and railway are newly added to the divisions supervising environment," "eco-supervising and measurement are stipulated in detail," "ecological environment preservation is stipulated in detail," "restriction on the total emission volume is enhanced and stipulated in detail," "environmental assessment is suspended in the key area and basin where the environmental quality criteria has not been achieved, and the area under the key restriction standard on the total emission," "the legislation of the rural environment preservation is enhanced," "disclosure of information is enhanced," "the liability of accomplishment of environmental preservation and the assessment system for the government, environmental protection divisions, and their officers are provided," and "the status of accomplishment of the preservation target to the Standing Committee is reported." (Kitagawa 2013: page 335)

[6]Refer to the 5-year legislation plan promulgated in the Standing Committee (人大常委会公布五年立法规划) (The National People's Congress Net (中国人大网) (http://www.npc.gov.cn/npc/xinwen/syxw/2013-10/31/content_1812095.htm) 2014.6.1.

Subsequently, a major amendment was added to the draft. This became the second revision draft for proposal of the environmental protection law. The second draft was published in July 2013, and the hearing of public opinion was held until August 18. Based on the hearing, the context of the draft has been enhanced and was passed to the Standing Committee on April 24, 2014, and enacted on January 1, 2015. The extraordinary rapid amendment process was spurred by the environmental effects of various tangible environmental issues such as PM2.5 pollution problem. The contents and the characteristics of the revised law were drafted based on comparison with the previous version with special attention being paid to legal amendment and new provisions.

For example, the provision, "target" and the part, "promoting modernized establishment of socialism" were deleted from the previous draft and the terms "promoting ecological civilization improvement and facilitating sustainable economic and social sustainable development" were added to the new draft. "The protection of environment is a basic national policy for China," which has already been formerly widely recognized, was specified (Article 4). In relation to "environmental education and preservation," the roles of educational departments, schools, and the media were legislated as "educational departments and schools shall incorporate environmental protection knowledge into the curriculum of school education so as to cultivate the environmental protection awareness among students" and "news media shall carry out the publicity of environmental protection laws, regulations and knowledge, and facilitate the exercise of public supervision on environmental violation activities."(Clause 2 and 3, Article 9).

In addition, "the environmental protection plan," whose contents includes "objectives, tasks and safeguarding measures, etc. for ecological environmental protection and environmental pollution prevention and control" and "shall align with the planning on main functional zones, overall land use, and urban and rural development," is newly established to keep consistency with other development plans" (Clause 4, Article 13). In association with "environmental impact assessment" in the construction projects, the law explicitly states that "construction projects without environmental impact assessment developed in accordance with the law shall not commence to construct". This new provision aims to nip in the bud of many rampant completed construction cases which failed to obtain assessment approval before project commencement. The new amendment also provides that prior approval of environmental impact assessment for development projects is required before work commencement (Clause 2, Article 19). For those who violate these provisions, the authority concerned "shall order them to stop the construction, impose fine penalty, and may require restoration of the construction sites" (Article 61).

To make environmental supervision responsibilities clear, "the State adopts environmental protection target accountability and performance evaluation system. The fulfillment of environmental protection target for People's governments at federal and the local level" shall be incorporated as an appraisal criterion into the performance evaluation system for departments with environmental supervision responsibilities at the same level of government and other persons responsible, and the evaluation results shall be made public. The fulfillment of an environmental protection target above county government level shall be incorporated as an appraisal criterion into the performance evaluation system for departments with environmental supervision responsibilities at the same level of government and other persons responsible and form an important basis of the evaluation. The evaluation results shall be made public (Article 26). Furthermore, "the people's governments at or above the county level shall report, on an annual basis, the environmental conditions and the completion of environmental protection targets to the people's congress at the same level or its standing committee," and major environmental incidents shall be reported promptly to them as well (Article 27). This is to enhance supervision by the people's congress.

Natural resources legislated as " Exploitation and utilization of natural resources shall be developed in a rational way that conserves biological diversity and safeguards ecological security" (Clause 1, Article 30), "for introduction of exotic species as well as the research, development and utilization of biotechnology, effective measures" must be given "real teeth" for the prevention of the destruction of biological diversity (Clause 2, Article 30). As a result, the establishment of an ecological protection compensation mechanism and the state's provision of guidance for the local people's governments of beneficiary areas and ecological protected areas on ecological compensation through consultation or market rules (Article 31) are legislated. This seems to reflect the environmental initiative to establish ecological protection compensation regulation whose draft is being made out at present.

Regarding "wastes and recycling, "the government departments and other institutions shall give priority to procure and use energy-efficient, water-efficient, and material-efficient products, equipment, and facilities (Clause 2, Article 36), and local people and governments are required to undertake "measures to organize the sorting and separation, as well as the recycling of municipal solid waste" and citizens' duty to sort and separate municipal solid waste are regulated (Article 37, 38).

"A total emissions control quota" is legislated, and regarding regions that fail to fulfill the total emissions control quota or achieve environmental quality targets, EIA approval shall be suspended for their new construction projects (Clause 2, Article 44) in order to enhance its effectiveness. Environmental protection of farmland from pesticides, chemical fertilizers, and other unsustainable cultivating pratices is legislated and is the responsibility of the people's government at various county levels (Article 49).

In this revision, the "information disclosure and public participation" and the "legal liability" are particularly worthy of attention. On the issues of "information disclosure and public participation," the competent department of environmental protection administration under the State Council at or above the provincial levels shall regularly publish environmental status bulletins (Clause 1, Article 54). The promotion of environmental information disclosure at the local level is expected. Regarding the public participation in the process for environmental impact reports, it is explicitly stated as follows: "the project owner of a construction project for which an environmental impact report should be prepared pursuant to the law shall explain relevant situations to the potentially affected public when preparing the environmental impact report and solicit public opinions." After the receipt of reports by the competent department, they shall make public the full text of the environmental impact reports except in cases of commercial secrets and confidential circumstances. The construction projects are regulated to solicit sufficient public comments to make it more concrete (Article 56). And regarding public lawsuits, it is stated clearly that "for activities that cause environmental pollution, ecological damage and public interest harm" (Article 58), social organizations that have their registration at the civil affair departments of people's governments at or above the municipal level with subdistricts in accordance with the law and specialize in environmental protection public interest activities for five consecutive years or more and have no law violation records may file litigation in the people's courts. More than 300 NGOs are qualified nationwide. There are environmental protection courts established in Wuxi, Guizhou, and Yunnan, where case[7] litigation filed by the (All-China Environment Federation) 中华环保联合会 was made in favor of it. However, it is acknowledged that the situation should be monitored closely.

In relations to the provisions of "legal liability" and "daily-basis fine" which the voters strongly supported, they were legislated as follows: "where an enterprise, public institution or other producer or business operator is fined due to illegal discharge of pollutants and is ordered to make a correction, and if the said entity refuses to make a correction, the administrative organ that makes the punishment decision pursuant to the law may impose the fine thereon consecutively on a daily basis according to the original amount of the fine, starting from the second day of the date of ordered correction" (Article 59). In addition to the basic legislation, the categorization of activities as illegal acts has been widened according to each local situation. Furthermore, where a construction project of a stakeholder has not been subject to environmental impact assessment in accordance with the law and the stakeholder concerned is ordered to stop construction but refuses to comply, or where pollutants are discharged, the case shall be transferred to the public security authority by the competent environmental protection administration of the people's government, and the person directly in charge and other personnel subject to direct liabilities shall be imposed a detention of 10–15 days (Article 63). Seeking liability of units are enhanced as "for key pollutant-discharging units that fail to disclose at all or disclose false environmental information, environmental departments of government at or above county level shall order them to disclose such information, impose fine penalty, and issue public notice on their violation" (Article 62).

[7]On December 15, 2003, the judgment on the case of secret disposal of contaminated wastewater into the Nanming River by GuizhouDingba Paper Factory passed by Qingzhen People's environmental prevention courts was made in favor of All-China Environment Federation to accept the request for suspension of disposing contamination. This was the first winning lawsuit in the history of public lawsuits (Reported by Ma Yong, Director of Lawsuit Division, All-China Environment Federation, in Beijing, September 2012).

From the above, the following points can be considered as the salient characteristics of the proposal: ① The reach of the existing regulation centered on environmental pollution is expanded to the field of the protection of nature and the prevention of waste; ② information disclosure by the government and corporations is required; ③ supervision and its function for residents and media are attached greater importance and their participation is empowered; and ④ legal responsibility is enhanced. In particular, ④ is enhanced greatly compared to the second revised draft of environmental protection law proposal—an indication of strong determination of the legislators. Before its enforcement, the contents of the proposal were widely made known to the people who were eager to see it coming into force in 2015. However, the difficulty of balancing environmental protection with economic development has hampered effective implementation locally. Also, the inconsistencies with other environmental laws which remained unchanged has been considered as another stumbling block that impedes effective implementation. In addressing these impediments, it is necessary to embark on an institutional reform to enhance governance machineries—the key to effective administration of environmental laws and policy administration.

1.4 Law Administration and Governance

Various controversial issues on the existing Chinese administrative law enforcement have been pointed out. These include "the liability entities of the law administration are vague," "the enforcement abilities of the environmental administration are weak," "underdevelopment of law administration," "low-quality human resources of law administration staff," "the process and the measurement of law administration are opaque," "the supervision system for law administration and the legal liability system are defective," etc. (Wang and Qin 2003: pages 144–150). In the background of these controversies, other issues have been simultaneously identified. These include the details of environmental protection law which are not distinctively clarified. Also, the law contains many sections whose meanings depend on the interpretation and decisions by the administrations, such as that the environmental protection law, whose details have not been clarified, contains many portions that depend on interpretation and decisions by the administrations. To compound the problem, the Chinese have a unique law culture where people regard the law as given by the upper hierarchy with the consequence that many tend to resort to law evasion.

"The environmental prevention administration divisions, State Environmental Protection Administration of China (abolished in 1998), and National People's Congress Environment Protection and Resources Conservation Committee join the constitution of environmental policy," "authorities of administration and supervision overlap," "integration between administrative regulations and laws," "authority to give their original interpretation on the process of law administration and supervision by courts and the Public Prosecutor," "expenses for local courts at each level are provided by local governments," and "local protectionism" are also pointed out as some of the controversial issues (Bao 2009: pages 55–71).

Based on the author's recent research and study, in order to prevent environmental pollution and destruction of the ecological environment, to protect society's environmental interests, and to improve environmental governance, it is necessary to consider the following five points[8](a) the establishment of a supervision system for law enforcement or policy implementation by the administrative and financial divisions in the local governments, (b) to reform the executive personnel management system, (c) to reform local governments' finances, (d) to reform environmental administrative divisions, and (e) to expand information disclosure and actual public participation (Kitagawa 2012: page 30).

1.4.1 Establishment of a Supervision System

It can be difficult to fairly assess whether administrative and financial implementations proceed appropriately under social ethics in a regime where the leader from the Communist Party holds an additional important official position. Although there are currently administration inspections and an infringement punishment system in the Communist Party's regulations, a supervising function based on environmental protection as a matter of course is required to ensure effective enforcement. These include judicial improvement, and the establishment of a system such as a supervision system by residents like an ombudsman and an ex post facto supervision regime like citizen's suit in Japan is needed. For this establishment, administrative processes in ensuring transparency and information disclosures are necessary.

1.4.2 Reform of the Executive Personnel Management System

An examination system including factors such as resources, the environment, and sustainable development should be incorporated into the assessment for government officials and leaders of the Communist Party. The Central Politburo of the Communist Party of China regulated "the achievement of the requirement of Scientific Outlook on Development (科学的发展观) and the trial measure of total performance assessment for leaders and leader-level staffs of the Communist Party" in 2006 and considered criteria for resource protection such as arable land and

[8]Grants-in-Aid for Scientific Research(B)・ Overseas Academic Research "Research for Improvement of Administration of the Environmental Law and Environmental Governance in Chinese Local Government"(2008–2010).

environmental protection. Furthermore, it is said that "Opinions regarding promotion of Scientific Outlook on Development and establishment of the assessment system for the leader group and the leader officers in the Communist Party" (Wang and Luo 2009) was regulated. Based on a local interview conducted by the author, it was revelaled that the actual situation has not been disclosed and hence, to what extent environmental reform has been implemented in actuality cannot be determined. In another interview conducted with an officer (a member of the Communist Party) from Shaanxi (陕西) local government on issue concerning "the regulation for the assessment of the leader group and the leader officers in the Communist Party (trial)," it was disclosed that environmental achievements as reflected in the assessment of the leading officers could not be ascertained.[9]

1.4.3 Reform of Local Governments' Finances

The work in the local governments in China follows plans and policies enacted by the central government on a top-down basis. We need to understand that the local governments that put a lot of priority on their economy make every effort to secure financial resources for the achievement of these works through the introduction of the 1990s' tax-sharing system. The central government should consider better governance for the finances of local governments such as transferring of tax resources.

1.4.4 Reform of Environmental Administrative Divisions

In a questionnaire survey (Wang et al. 2009)[10] conducted by Professor Jin Wang from Peking University, with the officers from the environmental administrative division, many of them pointed out the necessity of establishing a vertical control system by the central environmental administrative division. They strongly felt the importance of receiving direct control and instructions from the Ministry of Environmental Protection in order to counteract interference by their local government. On the other hand, there is a concern that conflicts may emerge among other divisions. More than direct instruction by the Ministry of Environmental Protection, the central administrative division is expected to attempt comprehensive instructions for the policies and the reform of the administration of each division in the local governments, especially the establishment of a greater environmental

[9]July 27, 2013, Xi'an, Shaanxi

[10]372 officials in the environmental protection divisions at the provincial, city, and county levels were interviewed from May to December 2007. Ninety percent of the interviewees were officials from the environmental protection divisions and related institutes at the provincial and city level.

division, including coordination, water, forestry, and oceanic resources and the promotion of the policies (similar view・ Wang 2011). From this point of view, the activities of groups such as the Human Settlements and Environment Commission of Shenzhen Municipality (深圳市人居环境委员会) and the water resource preservation group in Hubei, which coordinate the relationships among the divisions, can serve as a model.

1.4.5 Expansion of Information Disclosure and Actual Public Participation

Unlimited disclosure of environmental information could stir up anxiety among members of society in countries with a huge population like China. On the other hand, the number of issues that need citizens' cooperation for a breakthrough, such as global warming, waste concerns, and automobile pollution, is increasing. It is also desirable that information in relation to health and safety be disclosed (van Rooij 2006) It is by no means easy to swiftly disclose environmental information in China, but it is necessary to share the such information across society as a whole as this will help to solve environmental issues and improve environmental governance. Indeed, information disclosure is a precondition for enhancing effective environmental governance. Although the Regulation of the People's Republic of China on the Disclosure of Government Information and Measures for the Disclosure of Environmental Information (for trial implementation) has been enforced since May 2008, there is still a need to monitor its changing situations continuously.

1.4.6 Others

1.4.6.1 Introduction of Sustainable Assessment

The introduction of sustainable assessment, which gives prior assessment to various factors such as the economy and society and evaluates plans and construction projects from the point of view of sustainable development, is needed rather than an environmental assessment emphasizing only environmental consideration. The institutional setting in China, where the strategic environmental assessment for plans is stipulated in the law and economic factors are heavily emphasized, could be better than the one in Japan. For administrative processes, ensuring transparency and actual public participation is necessary.

1.4.6.2 Development of Human Resources with Specialized Knowledge (Introduction of a Qualification System and Enforcement of Specialist Courses)

The improvement of specialized knowledge in the environmental field and the ability to solve issues by enforcement of human resources such as judges and the officials in the environmental administrative division and a qualification system in the environmental field with specialist courses are necessary. Simultaneously, the improvement of environmental consciousness of school students and adults is also important.

References

Wang, Jin, Zhang, An, Yan, Houfu, "Environmental Law Administration is Difficult. Where Difficulty lies?-A report of Questionnaire to staffs of Environmental Protection Department in China", Environmental Law in China, Law Press China, 2009.

Wang, Jin "Eco-Nomocracy - Did we succeed?" Peking University Press, August 2011.

Wang, Jin (Arthor), Kou, Xin (Translation), Kitagawa, Hideki (translation supervision) "On the Effective Methods to Partly Amend the Environmental Protection Law of P.R.C in China" Rryukoku Journal of Policy Science, Vol.1 No.2, March 2012.

Wang, Xi, Qin, Tianbao "Implementation of Pollution Control Law in 10 years after Rio", Ryukoku Law Review, December 2003.

Wang, Xi, Luo, Wenjun, "A Opinion about Causes of Environmental Management's Failure" Report of Workshop in Shanghai, September 7, 8, 2009.

Bao, Maohong, "Environmental Governance in China and Environmental Cooperation in North-East Asia", Harushobo, 2009.

Kitagawa, Hideki "Environmental Law and Policy and its Implementation of China" (from "Environmental Law and Policy and Governance of China" Introduction) Koyoshobo, January 2012.

Kitagawa, Hideki "Effects and Issues of Chinese Environmental Law's Implementation for 30 years-Mainly on the Amendment to the Environmental Protection Law and the Environmental Dispute Institution" (from chapter 12 of Hideki Kitagawa et al. edition "Development and Change of Modern Chinese Law", Seibundou) July 2013.

Koyanagi, Hideaki (2010) "China department of environmental issues" Sososha,April 2010.

Benjamin van Rooij "Regulating Land and Pollution in China-Lawmaking, Compliance and Enforcement, Theory and Cases" Leiden University press, 2006.

Chapter 2
Evolution of Environmental Thought and Enforcement of Environmental Protection Legislation in China: The Status Quo

Jin Wang

Abstract This paper analyzes the background, the process, and the evolution of environmental thought of the Standing Committee of the National People's Congress in its approval of the Environmental Protection Law (for trial implementation) in 1979. It also discusses its subsequent revision in 1989 followed by its eventual official enactment in 2014. It further evaluates the effectiveness of this legal enactment since it came into force in 2015 in addressing various pressing environmental issues confronting China today. It is concluded that despite the evolution of thought over the past few decades in giving priority to environmental protection to economic growth, casual conservation on the ground indicates that this does not necessarily transform into action in reality.

Keywords Environmental Protection Law • Revision • Legislative thought • Enforcement of law • Existing problems

Abbreviations

CPCCC	The Communist Party of China Central Committee
CRAERL	China's Research Association of Environment and Resource Law
EIA	Environment impact assessment
EPL	Environmental Protection Law
LAOSC	Legislative Affairs Office of the State Council
LPCAP	Law on Prevention and Control of Atmospheric Pollution
MEP	Ministry of Environmental Protection
NESDP	National Economic and Social Development Planning
NPC	National People's Congress

J. Wang (✉)
Law School, Peking University, Beijing, China
e-mail: wangjin_law@pku.edu.cn

H. Kitagawa (ed.), *Environmental Policy and Governance in China*,
DOI 10.1007/978-4-431-56490-4_2

NPCERPC	National People's Congress Environment and Resources Protection Committee
PDUs	Pollutant Discharging Units
SC	State Council
SCNPC	Standing Committee of the National People's Congress
SDPC	State Development Planning Commission
SEPA	State Environmental Protection Administration
SPC	Supreme People's Court
SPP	Supreme People's Procurator
UNCED	United Nations Conference on Environment and Development
UNCHE	United Nations Conference on the Human Environment

2.1 The Creation of EPL and the Development of Environmental and Resource Protection Laws in China from 1979 to 2010

After the fourth deliberation, the Standing Committee of the National People's Congress (hereinafter referred to as NPC) of the PRC approved the newly revised Environmental Protection Law (hereinafter referred to as EPL) on 24 April 2014, which came into force on 1 January 2015. If the 1979 EPL (for trial implementation) enacted by the Standing Committee of the National People's Congress (hereinafter referred to as SCNPC) was the first milestone in Chinese environmental legislation history, the revised draft of the EPL approved on 25 April 2014 should be regarded as the second milestone in Chinese environmental legislation history. Since the leading ideology of the revision in 1989 was "environmental legislation cannot be too strict" and "it cannot increase the burden on economic development," the fundamental principle was established as "environmental protection should be coordinated with economic development," resulting in more principal, declaratory, and call-to-action articles but less obligatory, specific, and liability articles in national environmental legislation from the late twentieth century to the early twenty-first century. As a result, when applying laws to environmental illegal acts and when ascertaining where the environmental infringement responsibility lies, both official environmental agencies and pollution victims felt that it was difficult to find articles to punish illegal acts or to protect and remediate victims.

Although the sum of environmental and resource protection laws exceeds 10% of all the national laws in China, environmental pollution and ecological damage are still increasing, so that environmental and resource protection laws are called "soft laws." After a 25-year legislation plan of revision developed from "The Eighth NPC" to "The Twelfth NPC," after the aim of legislation changed from "limited modification" to "overall fulfillment" in three years, after the revision pattern changed from "draft amendment" in the first deliberation to "revised draft" in the third deliberation, the revised draft of the EPL was finally approved on 24 April 2014, after going through the fourth deliberation process of the SCNPC.

This paper will analyze the current problems in the enforcement of Chinese environmental protection laws, based on the review of several relevant legislations and modifications and the change of legislative thought.

2.1.1 *Birth of the EPL (for Trial Implementation) in 1979*

Since participating in the first United Nations Conference on the Human Environment (hereinafter referred to as UNCHE) in Stockholm in 1972, environmental protection has been regarded as an important approach to take when participating in international politics and economic cooperation in China.

The State Council (hereinafter referred to as SC) delegated the State Development Planning Commission (hereinafter referred to as SDPC) to hold the First National Environmental Protection Conference in Beijing in August 1973. The Conference enacted Several Issues on Environmental Protection and Improvement, clarifying a strategic guideline on environmental protection, that is, "Plan comprehensively, arrange reasonably, utilize systematically, turn harm into good, rely on people, act together, protect the environment, and bring benefits to the people." In 1974, the SC established an environmental protection group. Afterward, relevant departments of the SC issued environmental standards in succession, regulating specific environmental qualities and pollution emission standards. The NPC regulated in Article 11 in the third Constitution of the PRC (which was approved on 5 March 1978) that, "The State shall protect the environment and natural resources and prevent pollution and other public nuisances," while many countries just failed to accomplish the environmental rule of law as they advertised.

Actually, after the environmental enlightenment brought about by the 1972 UNCHE, the Chinese government began to find that air pollution in large- and medium-sized cities as well as industrial areas was serious, rivers, lakes and seas all over the nation were polluted to different degrees, the scope of groundwater pollution was increasing year by year, and natural environment was quite seriously damaged (Li 1979). Taking industrial pollution as an example, according to the statistics from ten provinces and cities like Beijing, Shanghai, Heilongjiang, Hebei, Henan, Anhui, Fujian, Liaoning, Sichuan, Shanxi, etc., in 1979, industrial and domestic waste water totaled 37.30 million tons per day, an increase of 7% compared to that in 1978 and 21% compared to 1977; waste gas totaled 6300 million cubic meters per day, an increase of 5% compared to 1978 and 17% compared to 1977; and waste residue totaled 310 million tons per year, an increase of 3% compared to 1978 and 10% compared to 1977 (Qu 1989).[1]

[1]Besides, the Keys of report of environmental protection work issued in October 1978 also summarized relevant environmental pollution situations. See State Environmental Protection Administration (hereinafter referred to as SEPA) of China, Literature Research Center of the CPC Central Committee (eds) (2011) Selected works of important documents on environmental

In November 1978, the speech named "Emancipating the mind, seeking truth from facts, and looking ahead in unity" made by Deng Xiaoping in the closing meeting of the Working Conference of the Central Committee of Communist Party of China (hereinafter referred to as CCCPC) mentioned specifically that such laws as the EPL must be enacted as soon as possible. Afterward, the document for this meeting issued by the CCCPC mentioned specifically that "it is necessary to enact laws and regulations on pollution elimination and environmental protection." For this purpose, the drafting work of the EPL began intensively under the lead of the SC Leading Group on Environmental Protection.

Following the legislative intent at that time, and compared to the USA, Japan, and other nations under the rule of law, the drafting group thought that the EPL was the fundamental law of national environmental protection work, so one of the purposes of enacting the EPL was to determine the fundamental principles and policies on national environmental protection in legal terms, and the other was to determine the future blueprint for establishing the legal system for national environmental and resource protection. Therefore, the preliminary design for the legislation on environmental protection was "only laying down the fundamental principles and policies on national environmental protection, and leaving specific stipulations for the future formulation of specific regulations and detailed rules of air protection laws, water protection laws, etc." (Li 1979).

Although the CPC and national leaders had recognized the importance of enacting the EPL from a strategic height, many members of the NPC and government officers believed that the economy had not been growing and had zero stock[2] in China. Many opposite opinions that it was too early for environmental protection legislation, which might affect the rapid development of the industrial economy, appeared when the EPL draft was first submitted for discussion in the SCNPC. Under these circumstances, the SCNPC approved the first EPL "in principle" and "for trial implementation."[3] It was quite striking, given that the national legal system building had just been started, and the total number of laws was just over ten.

The implementation of the EPL (for trial implementation) marked that the environmental protection cause in China had begun to head down a legal track. This EPL first clearly established the fundamental principles, missions, and policies as well as the object of environmental protection in China; clarified fundamental

protection in the new period. Central Party Literature Press, China Environmental Science Press, Beijing, p 5-8.

[2]"Stock" in the economy means the status of assets and liabilities for a given period of time or assets and liabilities held for a given period of time.

[3]"Approving in principle" means a way to approve a law draft when the SCNPC agrees on the fundamental spirit of the law but does not completely agree on all specific articles. "For trial implementation" means although the SCNPC thought the law draft was not sufficient or the legislation condition was not mature, law was needed in real life; therefore, the law was approved. It was a custom in legislation by the NPC and the SC before the 1980s. "Interim" is a similar concept with "for trial implementation."

principles of environmental protection like "taking environmental protection into overall consideration and arrangement," "putting the emphasis on prevention, combining prevention with treatment and conducting comprehensive treatment," and "those who cause pollution are responsible for treatment," etc.; and established a series of such fundamental rules for environmental protection as environment impact assessment (hereinafter referred to as EIA), the "Three Simultaneities" system, "collection of pollution charges," etc. After the implementation of this EPL, special environment agencies were established in the central government, provincial governments, municipal governments, and parts of county governments, as well as in industrial and agricultural management agencies and many large- and medium-sized enterprises, which laid the administrative institutional foundation for the enforcement of EPLs. To understand the situation of environmental protection in large- and medium-sized enterprises, in 1980, the SC Leading Group on Environmental Protection organized large- and medium-sized enterprises to fill in a Chart of the Investigation of the Status Quo of Environmental Protection in Large and Medium Enterprises and produced a statistical summary. This was the first batch of information that the enforcement of EPLs in China relied on.

In the Fourth Session of the Fifth NPC held in December 1981, the former president of the SC, Zhao Ziyang, first required according to the EPL (for trial implementation) that "it should prevent and control pollution and protect ecological balance" in the "ten principles of national economic development" in the Government Work Report. In November 1982, environmental protection was first adopted in the "plan of social development" in the "Sixth Five-Year Plan for National Economic and Social Development of the PRC," which was approved by the Fifth NPC.

However, China's economic system reform began in the 1980s. Due to the "approving in principle" and "for trial implementation" arrangement, officers from all the departments of the SC and local governments at all levels generally considered that the law should be flexibly implemented based on the need for economic development. For example, the method of implementation is mainly for large- and medium-sized state-owned industrial and mining enterprises. The environmental protection department did not directly supervise or manage the enterprises' discharge of pollutants under the planned economy but required these enterprises to reduce emissions in combination with technical reform and under the guidance of various ministries, so as to gradually meet the national emission standards. Besides, the protection of the ecological environment was only emphasized after the development of natural resources was satisfied. Therefore, the enforcement of EPLs had little effect.

Taking the period of the "sixth five plans" on national economy and social development (1981–1985) as an example, under the condition that the production processes, technologies, and equipment of the enterprise were still quite poor, China's total output value of industry and agriculture grew at the rate of 11% per year on average, which was not only higher than that in all previous periods in China but also higher than that in many countries in the same period (SEPA (eds) 1988). Because of too much investing in fixed assets and overheated economic development, China's environmental problems were not solved; instead, China began to "become one of the countries where pollution was the most serious" (SEPA (eds) 1988).

2.1.2 The SCNPC Revised and Approved the EPL for the First Time in 1989

The Constitution was revised and approved by the NPC in 1982.[4] In order to adapt the modifications and changes to the national fundamental political systems by the Constitution, in 1983, the SC put the revision of the EPL (for trial implementation) into the 1982–1986 Economical Legislative Plan, which was approved by the SC in principle.

The following three factors were mainly considered in the revision of the EPL (for trial implementation): first, the legislative basis was changed due to the amendment of the Constitution in 1982; second, China's economic system transitioned from a planned economy to a planned commodity economy, and the national system of the EPL was far from perfect; and third, from the point of view of the effect of implementation, the 1979 EPL (for trial implementation) was less normative and less binding, many of its provisions were not appropriate, and some of the separate laws and administrative regulations were ahead of the EPL (for trial implementation) so that its function gradually weakened. For all these reasons, three guiding thoughts of this revision were established: first, this revision shall adhere to the principle that environmental protection should be coordinated with economic construction; second, the revision shall adhere to the normalization and binding force of the EPL; and third, this revision shall be in accordance with the mode of comprehensive and fundamental EPL. The revision work began during the second half of 1983–1989, which lasted for five years.

In 1983, the former Ministry of Urban and Rural Construction and Environmental Protection took the lead to organize a drafting group for revision of the EPL (for trial implementation). At that time, the main guiding ideology was this: adhering to the principle of seeking truth from facts, according to general principles, the total task of the Constitution and the socialist modernization construction; turning mature environmental protection policies into laws; and making environmental law more scientific, rich, perfect, and effective through the revision.

In early 1986, the draft for examination and approval of the revised draft of the EPL was submitted by the drafting group to the Bureau of Legislative Affairs under the SC, which was in charge of sending copies of the draft to various departments of the SC and local governments, as well as collecting opinions from them.

---

[4]Article 26 of Constitution in 1982 stipulates, "The State shall protect and improve living environment and ecological environment, prevent and control pollution and other public nuisance."

However, due to China's tradition of administration-dominated legislation and the system of environmental protection that combines unified management with division of labor with separate responsibilities, government departments did not adapt to the rapid expansion of the administrative power held by environmental protection departments; therefore, the former adopted a negative attitude toward strengthening environmental administrative management. Concentrating on the scope of the authorities of environmental protection departments, the regulating scope of laws, the environmental management system, and the decision of whether to implement laws scientifically and strictly or not, environmental protection departments, resource management departments, economic administrative departments, and state-owned enterprises had fierce debates.

The most obvious struggle of the administrative authority was the "riding on a horse to enclose" campaign[5] about expansion and anti-expansion of administrative authority on separate regulations about the environment and resources among all departments: on the one hand, environmental protection departments focused on perfecting separate regulations on pollution control; on the other hand, natural resources and economical departments focused on separate regulations on the environment and resources, such as ocean, water, forest, land, grassland, wild fauna and flora, agriculture and fisheries, mineral resources, etc., and approved them before the revision of the EPL. In addition, many local governments also worried about the damaging economic effects that could result from strengthening environmental protection.

As a result, during the consultation of the draft for examination and approval, the concept that "economic development cannot be hindered due to environmental protection" prevailed.

In August 1989, the SC deliberated and approved the revised draft of the EPL and submitted it to the SCNPC for deliberation. By now, the protective goals of the draft of the EPL had been limited from "legalizing mature environmental protection policies, making the EPL more scientific, rich, perfect and effective through revision" to "putting regulative emphasis centered on preventing and controlling environmental pollution and limiting the scope of authorities of environmental protection departments." The measures stipulated in the draft for examination and approval, which successfully aimed to regulate the discharge of pollutants of enterprises in many countries under the rule of law, such as a pollutant discharge license, the discharge of pollutants beyond prescribed standards, and strict collection of pollution charges, were also deleted during the deliberation because some leaders in the SCNPC considered that "they may bring economic burdens to enterprises" and "they are incompatible with China's actual conditions."

[5]The phenomenon of "riding on a horse to enclose" means a kind of behavior of land grabbing by including into one's fiefs the land running across through someone riding a horse. In the context, the phenomenon of "riding on a horse to enclose" can be used as a synonym for competition.

In December 1989, the SCNPC approved the amended EPL. Compared with the EPL (for trial implementation), the EPL amended in 1989 made much progress. For example, the 1989 EPL established the principle that "Environmental protection must coordinate with economic and social development," clarified the management system of environmental protection of "combining unified supervision with the division of responsibilities among departments," and also explicitly stipulated the environmental responsibilities of local governments that "Local governments at different levels must be responsible for the environmental quality within their respective jurisdictions." Furthermore, the 1989 EPL stipulated the measures on environmental administrative management, such as EIA, "Three Simultaneities" system, declaration and registration of the discharge of pollutants, collection of pollution charges, and treatment within a prescribed limit of time, etc., and perfected a system of legal responsibilities, including administrative responsibilities, and civil and criminal liabilities.

However, since the NPC did not clarify the relationship between the EPL and other relevant laws regarding application and legal effect, there have constantly been problems about how to apply the law when different rules appeared between the EPL, emerging separate regulations on pollution prevention and control, and civil, criminal, and procedural laws. Furthermore, due to the nature of the EPL as the fundamental law in environmental protection, especially the deficiency of a national strategy on environmental protection and provisions about restricting the government's decision-making behaviors, there were situations when implementing the EPL in which it would have been better to apply separate regulations concerning pollution control than to conduct administrative enforcement and it would have been better to apply the provisions from civil, criminal, and procedural laws than to hold judicial adjudication.

In April 1989, the SC held the third National Environmental Protection Conference. Based on summarizing the experiences of environmental protection management in China and aimed at problems and institutional dilemmas encountered during the revision of the EPL, this Conference put forward "Three Environmental Policies" and "Eight Management Institutions"[6] from the perspective of policies and strategies on government environmental management as a supplement to the inadequate response of environmental administrative management to the EPL.

---

[6]"Three Environmental Policies" mean that environmental protection should carry out the policy of putting prevention first, the policy of those who cause pollution responsible for treatment, and the policy of improving environmental protection and management. "Eight Management Institutions" mean institutional methods which include the EIA, the "Three Simultaneities" system, the collection of pollution charges, quantitative checks for comprehensive renovation of urban environments, an objective responsibility system for environmental protection, declaration and registration of pollutant discharge, pollutant discharge licenses, treatment within a prescribed limit of time, centralized control of pollution, etc.—The author notes.

2.1.3 The Revision of EPL from 1993 Until To Date: The Unending Process

In 1992, the 14th National Congress of the CPC moved forward to clearly establish a socialist market economy. The NPC approved amendments to the Constitution in 1993, revising "The State practices planned economy on the basis of socialist public ownership" to "The State practices socialist market economy" in Article 15 and added content stating that "The State strengthens economic legislation and improves macro-regulation and control."

In June 1992, the United Nations Conference on Environment and Development (UNCED) was held in Rio de Janeiro, Brazil, in which the "sustainable development" strategy was reinforced after being unveiled and popularized in 1987 following the publication of the Brundtland Report. Later, the Chinese government put forward Ten Countermeasures on the Environment and Development, as a national program of action on coordinating economic development and environmental protection.

In the spring of 1993, the newly-established Eighth NPC Environment and Resources Protection Committee (hereinafter referred to as NPCERPC) organized some officials, experts, and scholars to hold China's symposium on environmental legislative work. The symposium revealed that the NPCERPC would start the revision of the EPL in 1989 in accordance with its deficiency.

According to my summary, the deficiency of the EPL in 1989 mainly included the following four aspects. First, the EPL was enacted based on a planned commodity economy, which was inconsistent and uncoordinated with a socialist market economy. Second, the EPL was deliberated over and approved by the SCNPC rather than the NPC, without having the nature of a national fundamental law. Third, separate legislation on environmental and resource protection was not enough to ensure that the government considered the overall relationship between the environment and development at the time of specific planning and macroscopic decision-making. Fourth, the main provisions of the EPL have been reiterated or modified by later separate legislation on environmental and resource protection (Wang 2003).

Actually, all the state legislative, judicial, and administrative organs devote considerable attention to the revision of the EPL. According to my investigation, the legislative organs mainly focus on the status and proper functioning of the EPL in China's environmental and resource protection law system. The judicial organs focus on the mutual relationship between the EPL and other separate legislations regarding legal effect and legal application of environmental and resource protection. However, based on their administrative authority and different interests and needs, the focus of the government and relevant administrative departments at all levels is more widespread. Under the practical condition of "administrative-dominated legislation"[7] in China, the understanding of the position and function of the

[7]"Administrative-dominated legislation" refers to the legislations mostly enacted by administrative organs.

EPL by the governments and relevant administrative departments at all levels directly affects and determines the revision of the EPL.

In this context, not every party could reach a unanimous agreement on how to modify the EPL. For this propose, during the seventeen years from 1993 to 2010, although deputies to each NPC had a large number of proposals for the revision of the EPL, limited to several reasons and concepts, such as "great difficulty, different understanding, and unknown direction," "large disagreements in the Legislative Affairs Office of the SC (hereinafter referred to as LAOSC) and other departments," and "scattered theoretical claims and lack of depth in related theoretical research," the revision work of the EPL has always been a "research, revise, and enact" kind of legislative plan throughout various periods.

From the EPL revised from 1989 to 2010, the SCNPC enacted and revised almost 30 legislations on the environment, resources, energy, cleaner production, and the recycling economy, which occupied more than one-tenth of all laws enacted by the NPC and the SCNPC. Compared with the fast-growing and constantly-improving separate legislations on environmental and resource protection, every time separate legislation on environmental and resource protection was revised, the guiding significance of the EPL as the fundamental law in environmental protection would get smaller and smaller. Hence, by the end of the 1990s, most of the regulations of the 1989 EPL had been revised and replaced by later-enacted separate legislations on environmental and resource protection. In fact, the 1989 EPL already existed in name only.

Furthermore, regarding a national, important, fundamental legal system involving environmental protection, the NPC revised the Criminal Law and added a new chapter to stipulate "Crimes of Undermining the Environment and Resource Protection" in 1997, using nine articles to stipulate respectively fourteen kinds of crimes concerning damaging the environment and resource protection, such as the crime of major environmental pollution accidents; the crime of illegal dumping, piling up, and disposing of solid waste; the crime of importing solid waste without authorization; etc.; in February 2011, the Amendment (VIII) to the Criminal Law approved by the SCNPC revised the "crime of major environmental pollution accidents" to the "crime of environmental pollution" and reduced the constitutive elements of crimes of environmental pollution. The Property Law approved by the NPC in 2007 established the basic rules for adjacent relationships to environmental protection. In 2009, the Tort Liability Law approved by the SCNPC had a special chapter stipulating "environmental pollution liability." The Civil Procedural Law revised by the SCNPC in 2012 set up civil public interest litigation especially for environmental protection.

All the above revisions and changes concerning state environmental and resource protection laws and relevant environmental protection provisions regulated in criminal, civil, and civil procedural laws signify that the 1989 EPL is far behind China's socialist legal construction and behind the development of and need for environmental protection in both concept and institution, so the 1989 EPL must be revised on a larger scale.

2.2 The Approval of the Newly Revised EPL: The Evolutionary Process

During the period from the Eighth NPC in 1995 to the Eleventh NPC in 2011, there were 75 bills for revising the EPL from 2,353 deputies to the NPC and two delegations from Taiwan and Hainan. Since the Tenth NPC, there were 18 suggestions for revising the EPL from 101 deputies to the NPC. Some of the NPC deputies even proposed that, considering the situation in which the NPCERPC had been continuously conducting research but still had not taken action since it was established in 1993, part of the law should be revised (Wang 2011).

2.2.1 The Scope of EPL Revision Between 2011 and 2012: The Gradual Process

Since August 2008, after 15 years of "research on the revision of the EPL" by three terms of the NPC, the newly established NPCERPC ambitiously declared that the current NPC must finish the task of revising the EPL. At the end of 2010, the Eleventh NPCERPC offered the proposal of revising the EPL to the SCNPC (Wang 2011), and in January 2011, the NPCERPC formally entrusted the Ministry of Environmental Protection (hereinafter referred to as MEP) with drawing up the draft of the Amendment to the EPL.[8]

The basic orientation of this revision was "limited revision," and the basic principles established were "setting limited goals and focusing on the emphasis," "stabilizing the institution and responsibilities," and "strengthening duties and improving the system." With respect to the articles, there were mainly contents about the following eight aspects and about the revision and improvement of their relevant "legal liability" articles (hereinafter referred to as "eight plus one," that is, EIA, a collection of pollution charges, treatment within a prescribed limit of time, public environmental rights, environmental standards, environmental supervision, pollution prevention and coordination across administrative regions, and government's responsibility over environmental quality). It was required that the revision process should mainly concern the problem that some articles could not connect

[8]In Chinese legislative practice, the method of "amendment" is generally used in legal revision activities when there are few contents to be revised, and the style as well as the structure of laws is not changed. By contrast, the method of "revision" is mainly used in comprehensive revision events when the revised contents are huge in number.

with or adapt to separate environmental protection regulations, which were urgently to be solved by the current EPL, and eliminate the negative effects on local legislation and administrative enforcement because of the inconformity among different laws and regulations, so as to strengthen the validity of the practice of the EPL.

However, the ideas regarding the revision from the main leaders of the NPCERPC were different from those of the MEP and scholars on environmental law from the very beginning. The main leaders of the NPCERPC thought that the primary cause of problems with enforcing Chinese EPLs was not legislation but inadequate enforcement by environmental protection departments and failure to implement the responsibilities of the government in environmental protection. Therefore, "the key point of revision is to improve the operability and authority of laws" (Wang 2011). On the contrary, general scholars on environmental law and the MEP thought that, although there were enforcement problems in environmental protection by the government, the essence was that there were fewer compulsory norms and authorization norms authorized for enforcement by the legislation on environmental protection. Some important legal institutions were imperfect, and the regulation on legal sanctions was not strict. So, though it would be a limited revision, it should also guarantee strictness and concretization of the above regulations and strengthen the regulatory authority for environmental protection.

In August 2011, the MEP formally submitted the recommended draft of the Amendment to the EPL to the NPCERPC. In October 2011, based on research concerning the opinions of various departments and places, the NPCERPC put forward a newly revised draft of the Amendment to the EPL and modified and deleted a lot regarding strengthening enforcement authority in environmental protection and other new institutions and measures issued by the MEP in the recommended draft of the Amendment to the EPL.

The MEP did not recognize the recommended draft of the Amendment to the EPL proposed by the NPCERPC. The MEP thought that there was a great distance between the content of the draft of the Amendment proposed by the NPCERPC and the new situation and requirements of environmental protection work in the new period, and the MEP hoped that the NPCERPC could continue to organize the demonstration and perfect the draft.

After that, the principal leaders of the NPCERPC coordinated with the MEP several times and repeatedly became involved in law seminars organized by relevant sectors of social organizations in the academia of environmental law, hoping that the MEP would accept the opinions of the NPCERPC. By the end of 2011, the NPCERPC examined and approved the Amendment to the EPL (draft) (hereinafter referred to as the First Deliberation Draft) and submitted it to the SCNPC in April 2012 for its first review. Considering that there was a strong dissent between the LAOSC and the MEP over the First Deliberation Draft, the SCNPC did not arrange for the deliberation of the Draft in time. Not until August 2012 did the SCNPC for the first time deliberate the Draft. It is worth mentioning that the SCNPC deliberated and approved the revised Civil Procedural Law, which for the first time established civil public interest litigation based on environmental protection.

Compared with the original EPL, the First Deliberation Draft revised 28 articles involving 42 paragraphs, added four new articles of twelve paragraphs, and structurally divided the second chapter into two chapters, that is, "Environmental Management" and "Supervision and Inspection," so that the EPL was changed from its six original chapters to seven chapters in the First Deliberation Draft.

What changes were made in the First Deliberation Draft to the EPL? The important ones will be introduced below.

First, in the First Deliberation Draft, the fourth article of the original EPL was revised from "making environmental protection compatible with economic construction and social development" to "making economic construction and social development coordinated with environmental protection." Although this revision was progressive, the original provision "The planning for environmental protection formulated by the State must be incorporated into national economic and social development planning (hereinafter referred to as NESDP)" was changed to "According to the outline of the NESDP, the environmental protection administrative department along with the relevant departments of the SC draw up a national plan for environmental protection" in the Draft at the same time. This revision not only detached the planning for environmental protection from the "Five-Year Plan" but also obviously legalized the practice of environmental protection being restricted by economic development.

Second, the First Deliberation Draft modified the rules setting the established environmental standards as "connecting with the environmental protection goal," namely, it wanted to restrain scientific environmental standards with the people-oriented "environmental protection goal," the purpose of which was to allow environmental standards restricted by the needs of economic departments and enterprises.

Third, although the goal of revising the EPL emphasized "stabilizing the institution and responsibilities," in the First Deliberation Draft, the provisions in relation to the supervision over environmental protection did not reduce relevant departments but increased maritime, industrial, and information technology, housing and urban construction, and other authorities in charge involved in environmental protection.

Fourth, EIA, which actually had major defects and should have been strengthened, underwent almost no revision in the First Deliberation Draft. And little consideration was given to the requirements of the introduction of the "strategic EIA" system which had been called for by most of scholars and environmental protection departments.

Fifth, on the revision of the provisions of treatment within a prescribed time limit, the academic opinion was to delete the article and replace it with provisions of measures for consecutive daily penalties. However, in the First Deliberation Draft, there were only meaningless modifications based on the original provisions with no compelling force without adopting legal measures, which have been proven effective in practice.

Sixth, in the aspect of the provisions on the public's environmental rights and interests, which drew widespread attention from the society, the First Deliberation Draft did not undergo any modification or response. In accordance with the interpretation of Wang Guangtao, the Director of the NPCERPC, the provision on an enterprise information public system was established for "the basic requirement of truly protecting the rights and interests of the public" (Wang 2011). In addition, the NPCERPC also thought that the provision stipulating in Paragraph 3 of Article 34 in the Draft that "The citizens, legal persons or other organizations can apply to the people's governments at or above the county level and relevant administrative departments of environmental protection for environmental information in accordance with the Law" was also a basic provision on guaranteeing public environmental rights and interests (Wang 2012). It was clear that the Draft attached the exercise and the protection of public environmental rights and interests to the obligations of enterprises and government departments, while the demand of the public for proactively exercising environmental rights was not taken into account. However, the practice has proven that for both government departments and enterprises, the environmental information disclosure in accordance with the law was still very limited for a period of time before and after revising the EPL.[9]

Seventh, on the provisions on the government's responsibility over environmental quality, while all the parties admitted the nonfeasance by the government as a main cause for increasing environmental problems and put special emphasis on the government's responsibilities, the Draft unexpectedly made no changes to this article!

Eighth, on the revision of the provisions on legal responsibility, for which was originally asked that clear and strict stipulations be given on legal consequences, it is regrettable that, in the First Deliberation Draft, there were only a few words that were changed without any improvement and with no progress. On the provisions on environmental protection, those stipulations in civil and criminal legislations that were not for the purpose of environmental protection were even better than those in the First Deliberation Draft.

For example, the Property Law of 2007 designed a special article for environmental adjacent relations, and the Tort Liability Law of 2010 set a special chapter for "tort of environmental pollution," both of which were stipulations about civil liability. Under a back drop in which strict civil liability concerning the environment was established in the basic civil law and had become more and more strict,

[9]According to the report of the Environmental information disclosure: Arrangement during three years released by the Public Environment Research Center and the US Natural Resources Protection Committee, the environmental information disclosure system in China's 113 urban governments was evaluated, and the average score is 36, of which there were eleven cities scoring above the approve line (60 points), less than 10% of the total. See Environmental information disclosure: Arrandement during three years (2011). Available via the official website of Public Environment Research Center. http://www.ipe.org.cn//Upload/IPEReport/environmental information disclosure_arrangement during three years_evaluating results of PITIin 2011_page_01.jpg. Accessed 6 November 6, 2012.

the Draft should have been made to refine the matter of principle and abstraction that existed in relevant provisions of the civil legislation and provide provisions on the elements of civil liability, the methods of assuming responsibilities, administrative and judicial settlement of disputes, and evidence and causal relationships, so as to facilitate access to judicial or administrative remedies for victims of pollution. However, the Draft, which should have made a major contribution to the progress of China's basic civil legislation, actually had no echo. As another example, in terms of a civil procedure system, in August 2012, the newly revised Civil Procedural Law established a "environmental public interests litigation" clause, which provided that "With regards to environmental pollution, infringement of legitimate rights and interests of many consumers and other behaviors that harm the public interest, the authorities and related organizations prescribed by the law can initiate litigation at the people's court." In addition, the newly revised Civil Procedural Law also provided that "Every party can apply to the people's court for notifying people with expertise to appear in court and bring forward comments on expert opinions issued by the identifiers' specialized questions," which established the legitimacy of appearances by expert witnesses in the court in future environmental litigations. The Draft should have had specific responses to these revisions of the Civil Procedural Law to help the environmental public interest litigation being launched as soon as possible, but as a result, nothing relevant is mentioned in the Draft. In this regard, Director Wang Guangtao gave the explanation that "The EPL cannot use environmental public interests litigation to make money for lawyers, but it holds the responsibility of keeping environmental protection departments accountable."

According to this interpretation of legislative instructions, the revision this time focused on resolving the problem that the cost of breaking the law was low but the cost of abiding by the law was relative high and at the same time clearly pointed out that enterprises should shoulder the responsibility to not only reduce their discharge of pollutants but also remove the impact of the discharge of pollutants on the quality of the public environment. For this reason, the approach adopted in the Draft provided a system of responsibilities for environmental protection and clearly stipulated the responsibilities of environmental protection by pollutant-discharging units (hereinafter referred to as PDUs). It made monitoring data public in accordance with the law, and it made treatment within a prescribed limit of time fall under the supervision of the government, but there were no reinforced provisions on administrative punishment. Taking the implementation of consecutive daily penalties advocated by the MEP and scholars as an example, the advantages of this provision have been proven by countries under the rule of law, and even in two cities, Chongqing and Shenzhen, which have established consecutive daily penalties as the implementation of the measures for administrative punishment in the local rules and regulations on environmental protection. It is thought that it is a feasible and effective legal guarantee to promote strict enforcement of orders and prohibitions. However, the seemingly too simple explanation given by the NPCERPC on the reason why they did not adopt the provision is that "it is not already covered in the existing EPL and relevant separate laws, and relevant government departments have not formed a consensus."

After the SCNPC deliberated the First Deliberation Draft, they needed to ask the public for comments and suggestions on the Draft. In September 2012, China's Research Association of Environment and Resource Law (hereinafter referred to as CRAERL) held a senior seminar on revising the EPL at Peking University. After the seminar, the CRAERL put forward the Opinion Book on Amendments to the EPL (draft) and submitted it to Chairman Wu Bangguo of the SCNPC. The Opinion Book pointed out that, limited to the draft which had been released, the form, structure, content, and the legal provisions were all very immature, far from reaching the basic requirements for further discussion and for a second deliberation, so it was difficult to continue to revise the EPL on the basis of this Draft. And the Opinion Book recommended that the Legal Council of the SCNPC further investigate and consult and draft a draft amendment that truly meets the requirements of the times and in line with the Scientific Outlook on Development. Therefore, this First Deliberation Draft was not mature, and the CRAERL suggested further arguments before its next submission for deliberation.

According to statistics, during one month of collecting comments and suggestions from the public, the SCNPC received 11,748 views from 9,572 Internet users in total (Yu et al. 2014). Most of the views expressed dissatisfaction with the First Deliberation Draft and the requirements for redrafting a new amendment.

In this context, until the end of the Eleventh NPC in January 2013, the Amendment to the EPL (draft) was not further deliberated.

2.2.2 *The Scope of EPL Revision in 2013*

In January 2013, the 18th CCP National Congress opened, and at that time the specific requirements for the construction of an ecological civilization were put forward. In March 2013, the 12th NPC was established. According to the CPC, 18 deliberations on the new requirements for promoting the construction of an ecological civilization, and deliberation by former SCNPC members, the Legal Council of the 12th NPC reorganized a team to revise the First Deliberation Draft.

In the process of revising this draft, in view of various issues raised by the community, the Legal Council of the NPC did the following: first, defined the revision of the EPL as "the basic law in the environmental protection field"; second, revised and perfected the basic systems and main contents of environmental protection on a large scale; and third, heard the views of the MEP and scholars on environmental law and organized several major demonstration sessions. On this basis, the second Amendment to the EPL (draft) (hereinafter referred to as the "Second Deliberation Draft") was formed.

Compared with the First Deliberation Draft, the Second Deliberation Draft changed the main contents almost completely. On the whole, the Second Deliberation Draft made 43 revisions to the articles of the original EPL, added a new chapter named "Public Participation and Information Disclosure" to the whole structure, and changed the number of the articles from 47 in the original EPL to 59.

The main contents added in the Second Deliberation Draft were as follows:

First, protecting the environment is a fundamental national policy of the State; coordination between economic and social development and environmental protection is needed; and environmental protection must adhere to the principles of giving priority to protection, focusing on prevention, conducting comprehensive treatment, engaging the general public, and enforcing accountability for polluters.

Second, a plan on environmental protection must include the objectives, main tasks, and safeguards for ecological protection and pollution prevention and control and be connected with a major function zoning plan, a comprehensive plan on land use, and an urban and rural development plan, among others.

Third, the State must encourage research on environmental criteria; conducting environmental monitoring should comply with the monitoring standards. Monitoring institutions must use monitoring equipment up to the national standards and comply with the monitoring norms. Monitoring institutions and the persons in charge thereof must be held responsible for the truth and accuracy of the monitoring data.

Fourth, the State must establish a coordination mechanism across administrative regions for the joint prevention and control of environmental pollution and ecological disruption in key regions and valley areas.

Fifth, the State must apply an objective system of responsibilities and an evaluation and review system for environmental protection, and the people's government at or above the county level must report the environmental situation and the achievement of environmental protection objectives within the corresponding administrative are as to the Standing Committee of the People's Congress at the corresponding level; the relevant local people's governments in key regions and valley areas that fail to reach the national environmental quality standards must develop plans for reaching such standards during a specified period and adopt measures to reach such standards as scheduled.

Sixth, the State must establish and improve an ecological compensation mechanism.

Seventh, the illegal discharge of pollutants by installing underground pipelines, using seepage wells or pits, and conducting perfusion or by other means to avoid supervision must be prohibited; entities discharging pollutants under intensified supervision must install and use monitoring equipment in accordance with the relevant provisions of the state and the monitoring norms, continue monitoring the pollutants they discharge, and preserve the original monitoring records; the State must apply a total discharge volume control system to key pollutants; and the State must, according to the law, apply a system to the discharge of pollutants.

Eighth, in the aspect of legal liability, where any construction employer commences construction without submitting environmental impact assessment documents for its construction project according to the law, the department with environmental protection supervision and administration functions will order it to cease construction and impose a fine and may order restoration to its original state; for an act polluting the environment or causing ecological damage in violation of public interest, the All-China Environment Federation (ACEF) and environmental federations at the provincial level may institute an action in a people's court. In addition, for illegally discharging pollutants, consecutive daily penalties will be imposed, and the natural person who commits the offense will be punished with a public security punishment; where any damage is caused by environmental pollution, the tort-feasor shall assume tort liability.

In June 2013, the SCNPC held the third meeting for deliberating the Second Deliberation Draft. From July 19 to August 18, after collecting comments and suggestions from the public, the Second Deliberation Draft received a total of 2,434 comments from 822 people (Yu et al. 2014).

In October 2013, according to the views from all sides, the Legal Council of the NPC formed the Third Deliberation Draft.

Based on the Second Deliberation Draft, the contents newly added in the Third Deliberation Draft mainly include: the impact on the environment should be taken into consideration when setting economic and technical policies; environmental protection departments have the authority to seize or impound pollutant-discharging equipment; the range of the subject in environmental civil public interests litigation will be expanded; a physical punishment will be imposed on directly liable persons and executives for violations by the enterprise; and where service agencies employ trickery in conduct, the EIA shall assume both administrative penalties and civil liabilities.

In view of the revision, since the Second Deliberation Draft had deviated from the goal of "limited revision" established in 2011, the "Amendment to the EPL (draft)" was therefore changed to "Revision Draft of the EPL". In this regard, the Legal Council of the NPC believed that the social public attached great importance to and had intense expectations for environmental protection. Faced with the grim situation of environmental pollution, the Central Committee of the CPC paid a lot of attention to environmental protection. The relevant departments had taken a series of preventive control measures, and the effectiveness was obvious. "There are some prominent issues in current practices. For instance, some local governments ignored environmental protection on account of economic development, some enterprises merely pursued profits while disregarding social responsibilities, law enforcement lacked force, punishments were without any bite, the cost of breaking the law was too low, public awareness of environmental protection was not strong enough, and the intensity of propaganda about environmental protection education was weak. On these issues, the Draft made a number of specific provisions. For this reason, it still requires further in-depth study according to the advice and views from all parties of the SCNPC members to revise and improve the Draft after this deliberation."[10]

[10]See Report by the Law Committee of the NPC on the revision of main issues in the second deliberation draft, Beijing, 14 October 2013. In addition, the revised drafts of Chinese laws are usually put to vote after three deliberations by the SCNPC, and this report indicated that the revised draft of the EPL would continue to be revised and improved. There were two reasons for this: one was to wait for the new requirements of the Third Plenary Session of the 18th CPC Central Committee on ecological civilization construction and the other was to tell the whole nation that the EPL was about to be revised not slightly but significantly.

The Decision of the Central Committee of the CPC on Several Major Issues Concerning Comprehensively Deepening the Reform was deliberated and approved at the Third Plenary Session of the 18th CPC Central Committee in November 2013. The Decision first proposed, "Deepening ecological environmental management reform by focusing on building a beautiful China; accelerating the building of systems to promote ecological progress; improving institutions and mechanisms for developing geographical space; conserving resources and protecting the ecological environment; and promoting the harmonious development between humans and nature via forming a new pattern for modernization." While claiming specific requirements to promote the construction of the rule by law in China, in order to speed up the construction of an ecological civilization system, the Decision also proposed overall requirements that "In order to promote ecological progress, we must establish integrated institutions and implement the strictest source protection system, damage compensation system, and accountability system. We will also improve the treatment of the environment and ecological restoration systems and use the systems to protect the ecological environment."

Compared to the Third Deliberation Draft discussed before, these requirements of the Decision had more extensive content and were more stringent, which needed to be specified and institutionalized in the Fourth Deliberation Draft.

2.2.3 *The Scope of EPL Revision in 2014*

In early 2014, the Legal Council of the NPC organized scholars on environmental law to conduct a number of seminars and intensive discussion about the concept of the "environment," a pollutant discharge license system, the relevance and distinction between treatments within a prescribed limit of time and orders to make corrections within a time limit, the applicable methods of consecutive daily penalties, the subject of environmental public interests litigation, and the requirements of the Decision concerning an ecological civilization system and its relationship with the Revision Draft of the EPL. The Third Deliberation Draft was modified largely on this basis.

For instance, the Third Deliberation Draft, compared with the Second Deliberation Draft, increased the number of clauses from 59 to 65; the Third Deliberation Draft newly revised 9 clauses involving 10 paragraphs and added 6 clauses involving 8 paragraphs on the basis of the Second Deliberation Draft. The Fourth Deliberation Draft added clauses to 65 on the basis of the Third Deliberation Draft and added 7 clauses involving 14 paragraphs, the modification range of which covered all the clauses. The content of the Fourth Deliberation Draft, totaling 70 clauses in 6 chapters, was not finally determined until the fourth deliberation of the NPC on April 22.

The main contents of the Fourth Deliberation Draft included:

- Specifying the requirements for public participation
- Expressly providing the implementation of garbage classification
- Setting June 5 as the Environment Day
- Stipulating an ecological red lines protection system
- Establishing a licensing system for the discharge of pollutants
- Providing for the government establishment of the mechanism of a public warning about pollution
- Encouraging and organizing research on how environment quality impacts public health
- Recording information about environmental violations into social credit files
- Expanding the range of the subject of environmental public interest litigation to relevant social organizations legally registered by the civil affairs departments of the people's governments at or above the level of a districted city
- Clearly defining that environmental tort cases, including environmental pollution or ecological disruption, will apply to the relevant provisions of the Tort Liability Law

The Eighth Session of the 12th SCNPC approved the revised EPL on 24 April 2014, which came into force on 1 January 2015.

2.3 The Enactment of the Newly Revised Environmental Protection Law: The Problematic Issues

On 25 April 2014, I was invited to publish an article entitled "Another Milestone in the History of Environmental Legislation" in the People's Daily. In my opinion, the newly-revised EPL (hereinafter referred to as the new EPL) was a milestone; furthermore, it cautioned public authorities that "Only making laws is not enough. There are more than six months before the new EPL comes into force, and every department in the SC, particularly environmental protection departments, should commence with checking up on relevant regulations and standards as soon as possible and establish regulations on law enforcement and a standards system in accordance with the new EPL in good time. Governments and their supervisory departments on environmental protection at every level, enterprises and public institutions, and the general public should seriously study and understand the new EPL directly concerning public health and their own interests and prepare for the new EPL to come into force (Wang 2014)."

However, national authorities at all levels have demonstrated obvious insufficiency in legislation and administration concerning judging and dealing with new legislation, administration, and other issues that have arisen since the implementation of the new EPL as follows.

2.3.1 Current Status and Problems of Separate Environmental Protection Legislations

What problems still exist in environmental legal institutions since the new EPL came into effect?

First, environmental legal institutions overlap each other. Taking the pollutant discharge license and completion acceptance in the "Three Simultaneities" as an example, the contents that are checked are substantially similar, both including pollution prevention equipment and monitoring pollutant discharge. For PDUs, if there is a pollutant discharge license other than for the completion acceptance in the "Three Simultaneities," it involves setting up two kinds of licenses on the same behavior, and this means that administrative counterparts need to submit two applications. Article 16 of the Law on Prevention and Control of Atmospheric Pollution (hereinafter referred to as the LPCAP) provides that "Licenses for new construction projects, expansion or reconstruction projects that discharge atmospheric pollutants shall be issued by administrative departments on environmental protection in accordance with EIA documents and the acceptance of facilities for atmospheric pollution control." This indicates that pollutant discharge license is based on EIA documents and the completion acceptance of the "Three Simultaneities," so these two institutions might conduct repeat investigations.

Second, there is a lack of coordination among environmental legal institutions. Citing the total discharge volume control system and EIA system as an instance to illustrate, Article 44 of the EPL stipulates, "For areas that exceed the state's total discharge volume control indicators for key pollutants or that fail to achieve the environmental quality objectives determined by the state, the environmental protection administration departments of the people's governments at and above the provincial level shall suspend the procedures for approving the environmental impact assessment documents on construction projects in such areas that will increase the total discharge volume of key pollutants." Accordingly, we must adhere to "Total Quantity Approval" in EIA to restrict new pollutant discharge generating from the source. As preconditions of the EIA approval, the total quantity index is, nevertheless, too general at present owing to the lack of concrete bases and standardized operations.

Third, the environmental regulation system was not straightened out. The environmental regulation system consists of the following two aspects. For the external regulation system, environmental protection departments have encountered some controversy with meteorological departments and transportation departments over the atmospheric field and with water management departments over water pollution. In terms of the marine environment, there is an entanglement between environmental protection and marine departments. Altogether, unified supervision and management should be confirmed. It should be clearly defined by special provisions the ways and methods toward unified regulation and other measures of supervision, but the EPL has failed to make a breakthrough. For the internal regulation system, cross regulation and acts of their own free will exist in different environmental protection departments established based on different legal institutions, which exacerbates the contradictions and discordant situation among different environmental legal institutions.

Fourth, the EPL has loose contact with other separate EPLs. Taking the EIA system as an example, Article 56 of the new EPL states, "After receiving the environmental impact report for a construction project, the department approving the EIA documents for construction projects shall disclose the full text of the report, except the part involving any state secrets or trade secrets, and if it discovers that public opinions have not been fully solicited regarding the project, the department shall order the construction employer to solicit public opinions." However, the Law on EIA only provides that the construction units must, before submitting an environmental impact report of the construction project for examination and approval, seek opinions from relevant units, experts, and the general public by holding demonstration meetings, hearings, or any other means. Thus, it can be seen that there is no provision about approval authorities' duty to disclose the full text of the report (Article 21 of the Law on EIA, the 9th SCNPC, 2003). So the Law on EIA is lacking matching provisions in the aspect of EIA report disclosure with the new EPL (Jin and Wang 2015).

Fifth, on the issue of how to revise separate EPLs, there are greater challenges. Along with the ceaseless deepening of the construction of the ecological civilization in China and the presentation of the strategic layout of "Four Comprehensiveness"[11] proposed by the Central Committee of the CPC, the social community's expectations have been greatly heightened, and the requirements for the environmental rule of law have been greatly improved after the new EPL came into force. Since the EPL, as the fundamental law in environmental protection, has been revised too many times ahead of the existing separate EPLs, the primary and urgent issue is to solve the problem of the maladjustment between separate EPLs and the practical needs of social management.

Citing an instance with the revision of the LPCAP starting in 2008, since the draft of the LPCAP commenced before the EPL approved by the 9th SCNPC in April 2014, the Revision Draft of the LPCAP submitted to the 9th SCNPC for the first deliberation in October 2014 provided general provisions over atmospheric environmental standards, planning for compliance within a prescribed period, automatic monitoring equipment, reduction and clean utilization of coal, pollution prevention of fuel by motor vehicles and vessels, restriction or prohibition measures for the traffic of motor vehicles, prevention and control of dust pollution during construction and agricultural atmospheric pollution, increasing penalties on atmosphere environmental violations, etc. The provisions soared from the 66 original articles to 125 articles. Nevertheless, the Revision Draft of the LPCAP had two

[11]"Four Comprehensiveness" means the goal that "comprehensively deepening reform, comprehensively strengthening running the CPC strictly, comprehensively implementing the rule of law, and comprehensively building a moderately prosperous society" shall be achieved till 2020.

major flaws: one is simply repeating the provisions in the EPL and loosely connecting to the EPL and the other is not making targeted provisions according to the features of preventing and controlling atmospheric pollution.

In June 2015, the Second Deliberation Draft of the Revision Draft of the LPCAP was resubmitted to the 9th SCNPC. Although the problems existing in the First Deliberation Draft had been modified, there have still been calls for "delaying the Third Deliberation Draft", "remaking", or "revising in a significant way" over major deficiencies that existed in the Second Deliberation Draft, and some scholars on environmental law even appealed for the reorganization of the revision of the LPCAP (Liu 2015).

2.3.2 *Status and Problems of Environmental Administrative Enforcement*

It took more than eight months from the time when the new EPL was approved until its came into force.[12] In order to guarantee the smooth work of the enforcing operations of the EPL after it came into force, the MEP has instituted a series of supporting documents.

On the one hand, the MEP has developed a number of supporting measures in consideration of the new contents of the new EPL, since the new EPL provides a lot more stringent enforcement measures, such as seizure or impounding (Article 25), consecutive daily penalties (Article 59), restricting production or suspending business for rectification (Article 60), detention (Article 63), etc. Within this 8-month "buffer period," a number of supporting documents have consequently been released. For example, the MEP, the Ministry of Public Security, the Ministry of Industry and Information Technology, the Ministry of Agriculture, and the General Administration of Quality Supervision, Inspection and Quarantine have jointly issued the Interim Measures for the Transfer by Administrative Departments of Cases of Environmental Violations for Which the Penalty of Administrative Detention May Be Applied at the end of 2014, which is a supporting measure for the new EPL on "administrative detention applicable to serious environmental violations." The MEP also issued the Measures for Implementation by Competent Environmental Protection Departments of Consecutive Daily Penalties, the Measures for the Implementation of Seizure and Impounding by Competent Environmental Protection Departments, the Measures for the Implementation by Competent Environmental Protection Departments of Limiting and Halting Production for Remediation, and the Measures for the Disclosure of Environmental Information by Enterprises and Public Institutions in January 2015. These documents above relate to consecutive penalties, seizure or impounding, information disclosure, and other

[12]The newly revised EPL was approved on 24 April 2014 and came into force on 1 January 2015.

contents, while they stipulate the range of application, enforcement procedures, penalty terms, and other relevant provisions.

On the other hand, for the sake of guaranteeing the smooth operation of administrative enforcement of environmental protection, some supporting documents have also been published in a timely manner. For example, the General Office of the SC, in order to ensure the implementation in a comprehensive manner of the linkage between administrative enforcement and criminal justice and the launch of the inspection on environmental protection, issued the Notice on Strengthening Environmental Supervision and Law Enforcement in November 2014. The MEP has made a clear statement in the Key Points in the Work of National Environmental Supervision in 2015 that traces the chain of responsibility backward in typical cases on violations of environmental regulations and impeaching civil servants for offenses of misconduct in office if they should have perceived major environmental violations or reported or dealt with the problems. These provisions produce a certain amount of pressure on the subject of environmental law enforcement, which reflects that environmental law enforcement will be further strengthened.

In the practice of administrative enforcement of environmental protection, the fast-growing cases, after the new EPL was performed, center around discharging pollutants beyond the prescribed standards as well as falsifying or forging monitoring data knowingly or maliciously in order to avoid regulations, etc. The number of the cases of administrative detention transferred to public security organs is increasing rapidly (Li et al. 2015).

The MEP informed the media of the implementation of the new EPL and supporting measures in June 2015. The information revealed that in March and April, the number of cases of consecutive daily penalties was 515% of the number in January and February, while the number of cases of seizure and impounding was 125% of the number in January and February, the number of cases of restricting production or suspending business for rectification was 237% of the number in January and February, the number of cases transferred for administrative detention was 197% of the number in January and February, and the cases transferred for suspected crime on environmental pollution was 204% of the number in January and February. According to the investigation from January to April, there were 160 cases about consecutive daily penalties, amounting to 112.2951 million yuan and a maximum penalty of 15.8 million yuan. There were 1,186 cases of seizure and impounding, 698 cases of restricting production or suspending business for rectification, 437 cases transferred for administrative detention, and 429 cases transferred for suspected crimes of environmental pollution (MEP 2015).

Putting aside this data, conflicts gradually appeared during the process of administrative enforcement of environmental protection. Although the new EPL has made changes to many institutions, there is still a big gap in the understanding and implementation of these new rules. Many places do not fully understand the true meaning of the new EPL during the process of administrative enforcement of environmental protection, and they still stick to their previous practices. "Local policies" still take a leading position. PDUs are not yet aware of the severity of the new EPL. The phenomenon of "turning a blind eye" is still common, which is prominently demonstrated in the following ways:

First, the perception regarding environmental protection of the local administrative departments has not yet been fundamentally changed. Because the perception of political achievements with "GDP first" still exists, the "local protectionism" phenomenon is still relatively common. For example, in September 2014, when the media exposed a pollutant discharge incident in the Tengger Desert economic and technological development zone, the director of the bureau of environmental safety supervision was still guaranteeing that a desert sewage incident had not happened in an interview on his honor. Ultimately, the event attracted the intense scrutiny of national leaders and society as a whole, and the corresponding PDU and relevant administrative departments received due punishment.

Second, the abilities of administrative enforcement regarding environmental protection are quite weak. Not obtaining approval before construction and other illegal phenomena abound and regulatory dereliction of duty and corruption are hidden problems. A direct reason that led to these problems is the lack of enforcement capability. On the one hand, we have insufficient staff for law enforcement. At present, there are only 60,000 officers for environmental law enforcement, a relatively small number in the grassroots, and the number of officers of environmental law enforcement in some counties may not exceed ten; on the other hand, the ability to collect evidence is low. When administrative penalties are imposed on enterprises, officers of law enforcement do not have a strong awareness of evidence, and the problems are not very clear given the light penalties. With the penalties increasing, the legal savvy of enterprises is increasing, and the probability of administrative proceedings against environmental departments is also increasing, so the scientificity of measures for environmental enforcement is facing great challenges (Li et al. 2015).

Third, the administrative enforcement of environmental protection has still not attained a "new normalcy." According to the performance of enforcement of environmental protection, a large amount of work of administrative enforcement of environmental protection is executed under "high pressure" by superior departments, and the atmosphere of the environmental rule of law is not favorable enough. For instance, Linyi City in Shandong Province is among the first group of negotiated cities after the enforcement of the new EPL, and under severe stress, the city vigorously is putting forward a comprehensive renovation to its atmospheric pollution, so the regional environment and air quality have obviously improved. However, the "iron hand" of pollution control activity on the basis of related laws is called a "shock treatment"; behind powerful pollution control, suspending businesses for rectification leads to numerous unemployed workers, and problems of enterprises that are confronted with the risk of capital chain cracks have been exposed. The "iron hand" pollution control activity in Linyi City has raised a wave of discussion and extensive attention in our society.

Fourth, PDUs do not treat the new EPL seriously. Under the pressure of "unlimited punishment," some enterprises initiatively invest more on environmental protection; they accelerate their rate of rectifying and reforming environmental violations. But the phenomenon of PDUs not comprehending new environmental institutions still exists. For example, the Bureau of Environmental Protection of Xianyang City in Shaanxi Province imposed consecutive daily penalties on Shaanxi Carbonification Energy Co. Ltd., since the latter refused to rectify its discharge of pollutants beyond the prescribed standards, and the penalty cost was 15.8 million yuan.

With regard to the consecutive daily penalties imposed by the Bureau of Environmental Protection of Xianyang City, Shaanxi Carbonification Energy Co. Ltd., charged the Bureau on 10 June 2015. The enterprise held the opinion that the consecutive daily penalties imposed by the Bureau violated legal procedure were enforced improperly. The case reflected the inherent idea of PDUs about the low cost of breaking the law and the ignorance of PDUs on the new rules in the EPL.

2.3.3 Situation and Problems of Environmental Judicature

Compared to environmental legislation and administration, the situation of environmental judicature is more optimistic. In 2007, Guiyang City in Guizhou Province set up the first environmental protection tribunal at Qingzhen People's Court. By 2013, all national people's courts judged 4,093[13] cases about the environment and resources. In June 2014, the Supreme People's Court (hereinafter referred to as the SPC) set up an adjudication division for the environment and resources, which established a specialized organizational form of environmental judicature in China. By 2014, 22 provinces, cities, or municipalities in China founded 376 environmental protection tribunals, collegial panels, or circuit courts.

On 13 June 2013, the SPC and the Supreme People's Procurator (hereinafter referred to as the SPP) jointly published the Interpretation on Several Issues concerning the Application of Law in Handling Criminal Cases of Environmental Pollution. On 23 June 2014, the SPP published the Opinions on Fully Strengthening Environmental Resources Trial Work to Provide Powerful Judicial Safeguards for Promoting Eco-civilization Construction. On 6 January 2015, the SPC published

[13]The data is from the report named Judicial actuality and problems in the implementation of the new EPL, made by Han Deqiang, vice director of the Judicial Research Center of Environment and Resources in the SPC on China Annual Seminar of Environment and Resource Law on 18 July 2015. According to the statistics by Sun Youhai, there were 117,308 cases of administrative punishment decided by environmental protection departments at all levels in 2012, 160,625 petition cases, and 892,348 complaint cases of pollution accepted through the Internet and by telephone. The three groups of data add up to 1,170,279 cases. As a result, the small number of cases accepted by the court cannot prove that the environmental disputes are less; instead, it only proves that most disputes are resolved out of court in a nonlegal way, and environmental protection tribunals are far from being the main way to resolve environmental disputes. See Sun Youhai (2014) Analysis and suggestion on the problems on current trials on environment and resources. China Court, 17 September 2014.

the Interpretation on Several Issues concerning the Application of Law in Hearing Environmental Civil Public Interests Litigations. On 7 January 2015, the SPC, the Ministry of Civil Affairs, and the MEP jointly published the Notice on Implementing the Environmental Civil Public Interests Litigation System. On 1 June 2015, the SPC published the Interpretation on Several Issues Concerning the Application of Law in Hearing Environmental Tort Cases. These series of judicial interpretations and regulatory documents, from adhering to the top-level design, accurately grasping the overall direction of reform and innovation including the setting principles of trial institutions for environment and resources, the jurisdiction model and judicial mechanism, and encouraging us to cross the river by feeling for the stones, to giving full play to the local courts' enthusiasm and initiative, have provided valuable experiences and references from two aspects for the next-step reform and guidance for the work of environmental judicature.

However, environmental judicature is also faced with the following problems:

First, from the perspective of internal factors, besides the superficial problems of different names of environmental protection tribunals, environmental judicature is faced with a prominent problem of scarcity of judges who are qualified for the trial of environmental cases. As environmental cases often involve the identification and admission of science and technology evidence, consideration on public policy, judgment and interpretation on the risks and uncertainties in cases, and comprehensive or selective application in the trial of criminal law, civil law, and administrative law, there is a high demand for the professional skills of adjudicatory personnel.

Second, from the perspective of external conditions, adjudication divisions for environment and resources at all levels are facing the problem that a large number of environmental disputes cannot enter the judicial process. It is known that "one cannot make bricks without straw." In fact, on the establishment of organizations, the adjudication divisions for the environment and resources of the SPC are mainly composed of judges from the former Second Case Filling Chamber of the SPC, who have not concluded any environmental cases since its foundation, but now investigation and research have become a major part of their work.

Finally, on the level of social background, although there are an amazing number of environmental disputes in China each year, few will eventually enter the judicial process. Obviously, one reason is that local governments reduce trouble from environmental events to the minimum on account of safeguarding stability and political achievements,[14] but the reason that access to environmental litigations in the current legal system is too "narrow" should not be ignored.

[14]Taking the reform of the filing system in the people's courts as an example, cases should be filed and examined by the people's court at all levels in China before 2015, resulting in the rejection by the court of a large number of group litigations; cases that have an impact on people's livelihood and cases that have a big impact on one region and the petition of most environmental cases were rejected by the people's court so as not to be filed and entertained. The reform of the judicial system in 2015 will change it into a filing registration system, which means that the people's court should not refuse to accept lawsuits without a hearing, which guarantees the right of the parties to appeal.

Under the triplet backgrounds, though the establishment of adjudication divisions for the environment and resources as the horn of the specialization of environment judicature has sounded, it is still necessary for the courts to strengthen the construction of professional skills and provide special and systematic theory-carding to the work of trials on environmental cases. At the same time, it needs to speed up the introduction of supporting policies and legislations to safeguard the specialization of environmental judicature and expand concrete practices of hearing environmental cases when the time is right.

2.4 Conclusion

It can be inferred from the revision of the principle of coordinated development[15] in the new EPL that the evolution in thought in China's environmental legislations has experienced a fundamental transformation from putting economic development first to giving priority to environmental protection. The reasons for such transformation heavily depend on China's severe reality of environmental pollution and ecological damage as well as the political background of constructing the ecological civilization proposed in the Third Plenary Session of the 18th Central Committee of the CPC.

However, China is not a state under the rule of law. The transformation of thought in environmental legislation does not necessarily mean the enforcement of law will also transform with this shift in thought to adapt to the EPL. In spite of the transition of the economic growth pattern required by the SC, most enterprises still make their profits based on the consumption of resources and energy as well as polluting the environment instead of technological progress. During the transition period, the burden of energy conservation and the reduction of discharge by enterprises must be reflected in those powerful officials in local CPC and government organs who have planned for the GDP rapid growth.

However, a written order made by Dalian Maritime Court in July 2015 obviously made the mistake in applying the EPL that the Court dismissed the claims of the nongovernmental environmental organizations on account that the EPL was not applicable to the nongovernmental environmental organizations in the case which nongovernmental environmental organizations sued the PetroChina for pollution by marine oil. To avoid plaintiffs' appeal, in the next day after the written verdict was made, Dalian municipal government convened both the plaintiff and the defendant to make an agreement by administrative settlement, which PetroChina invested 200 million yuan to establish a special fund for marine ecology for Dalian marine environmental protection and ecological restoration, while the plaintiff abandoned the appeal. The ultimate goal of this action is to avoid that this case continued to move forward in judicial proceedings.

[15]Article 4 in the new EPL stipulated, "To ensure the coordination of economic and social development and environmental protection," which is revised from "To coordinate the work of environmental protection with economic construction and social development" stated in the 1989 EPL. This revision is similar to that of Basic Law of Pollution Countermeasures of Japan in 1970, in which the provision of "the preservation of living environment should be coordinated with the healthy development of the economy" was removed.

Faced with the current economic downturn in China, the local CPC and government officers will inevitably intervene in the enforcement of law by environmental protection departments and the judgment by the people's court, so as to maintain the achievement of economic growth. After all, the current evaluation system for party and government leading cadres, which stresses economic statistics and indicators, has not changed fundamentally. Chinese reform on the judicial system is just getting started, but the phenomenon of local CPC and government leaders restricting or improperly intervening in the judicial organs' power of independent exercise over jurisdiction still exists. Moreover, due to the complex setting of administrative institutions related to environmental protection and regulation inside the government and severe crossing and conflicts of administrative power, dereliction of duty with regard to environmental regulation can be found everywhere. Though the environmental protection tribunals in the people's court at various levels keeps increasing, substantial results have not yet been promoted as expected. There is no doubt that the revision of separate EPLs is still on the front burner to basically guarantee the complete implementation of the new EPL.

However, it is worth mentioning that the central leading group for overall deepening reform examined and approved documents including the Scheme of Environmental Protection Supervision (for trial implementation), the Network Construction Plan for Ecological Environment Monitoring, the Pilot Program About Developing Off-Office Auditing of Leading Cadres' Natural Resource Assets, and the Measurement on Eco-environmental Damage Accountability of the CPC and Government Leading Cadres (for trial implementation) on 5 July 2014, which creatively put forward the concept of the co-responsibility of the CPC and the government for environmental protection, off-office auditing of leading cadres' natural resource assets, and lifelong accountability of eco-environmental damages. This action indicates the determination and confidence of the Central Committee of the CPC in enforcing ecological construction and also releases clear and definite policy signals and system guidance to the public.

References

Li, Chaobo (1979) Explanation on the draft of 1979 EPL (for trial implementation), 11 September 1979

Qu, Geping (1989) Environmental problems and resolutions in China. China Environmental Science Press, Beijing, p 123

SEPA (eds) (1988) China environmental protection cause (1981-1985). China Environmental Science Press, Beijing, p 2, 16

Wang, Jin (2003) On the present situation and revising orientation of China's EPL. Environmental Protection, vol. 6

Wang, Guangtao (2011) Revising the EPL well to adapt to the new times. China Environmental News, 3 November 2011

Wang, Guangtao (2012) On the amendment to the EPL (draft) in the twenty-eighth session of the eleventh SCNPC, 27 August 2012

Yu, Xiaojie, Gu, Ruizhen, Wu, Jingjing (2014) From the modified to amendments to the law on environmental protection after the trial of four crafted significant public. Available via Xinhua. http://www.xinhuanet.com/. Accessed 6 November 2012

Wang, Jin (2014) Another milestone in the history of environmental legislation. People's Daily, 25 April 2014

Jin, Hai, Wang, Yingfang (2015) Reform of EIA approval system: Progress, problems and countermeasures. Environmental Protection, 10: 36

Liu, Yiman (2015) Revising the LPCAP: Why scholars appeal to significant revision? Southern Metropolis Daily, 22 July 2015

Li, Penghui et al (2015) The new EPL shows deterrent force, the enforcement is facing new challenges——Dialogue with Ji Gang, Director of the Penalty Office, Environmental Supervisory Bureau of the MEP. World Environment, 3: 23

MEP (2015). http://www.mep.gov.cn/gkml/hbb/qt/201506/t20150615_303569.htm. Accessed 14 July 2015

Chapter 3
Land Contamination Legislation in China: The Emerging Challenges

Xiaobo Zhao

Abstract This paper examines the Chinese system and its problems in managing contaminated land. China is facing deteriorated land contamination and the management of contaminated land is now assuming greater attention in Chinese debates. A national framework on contaminated land management has been under preparation. Over the past decades, environmental laws and regulations have been developed in China to address contaminated land issues. A number of different mechanisms have therefore been established. However, the existing management system appears to be ineffective. By examining the practices of some leading nations, e.g. the USA and UK, some suggestions are provided which might be helpful for China to improve its management of contaminated land. The author argues that a systematic contaminated land remediation mechanism, a sounder liability scheme and a more effective financial assistance scheme should be considered by the Chinese legislature in developing a national regime for contaminated land.

Keywords Contaminated land • Land/soil contamination • China • Contaminated land remediation • Environmental liability insurance • Environmental law • Soil pollution

Abbreviations

CERCLA	Comprehensive Environmental Response, Compensation, and Liability Act
China MEP	The Ministry of Environmental Protection of China
EIA	Environment Impact assessment
EPB	Environmental Protection Bureau
NSBC	National Statistics Bureau of China
NSP	Nonpoint Source Pollution
SEPA	The State Environmental Protection Administration

X. Zhao (✉)
School of Law and Justice, University of Southern Queensland, Toowoomba, Australia
e-mail: bob.zhao@usq.edu.au

H. Kitagawa (ed.), *Environmental Policy and Governance in China*,
DOI 10.1007/978-4-431-56490-4_3

3.1 Introduction

Contaminated land is a site where toxic chemicals that have the potential to be harmful to the environment, and ultimately to humans, are present in higher concentrations than those found normally in the area. Contamination can occur as a result of industrial, agricultural or commercial activities (Chen 2005). It may also occur following waste disposal, spills or movement of contaminants from adjacent properties. Chemicals that could be found at a contaminated site include metals and metalloids (such as lead, cadmium, mercury or arsenic from mining tailings or landfill sites), oil, tar, harmful gases, solvents, pesticides or other hazardous wastes. Humans can be exposed to contaminants from a contaminated site via various routes. The actual degree of risk associated with a hazardous chemical often requires a site-specific risk assessment (Wong and Bradshaw 2002).

Land contamination has become a deep-seated concern in China. Frequently reported food safety crises and heavy metal pollution cases in China have deepened the spreading fears among the Chinese people. Take the cadmium-contaminated rice, for example. It was first reported in South China's Guangdong Province in 2013, but it has been found in a number of inland cities and found served at people's dining tables (Wong 2013). As estimated by the Ministry of Environmental Protection of China (China MEP hereafter), one-sixth of the arable land—about 50 million acres—suffers from land contamination/soil pollution (China MEP 2013). More than 8 million acres of China's farmlands are so polluted that they are no longer suitable for planting crops (China MEP 2013). At the same time, more than 13 million tons of crops harvested each year are contaminated with heavy metals, and 22 million acres of farmland are affected by pesticides (China MEP 2013). This is a clear reminder that land contamination or soil pollution, which has been continuously accumulating over the past decades, is eroding the foundation of the country's food safety and becoming a real threat to public health. More importantly, China has 22 % of the global population but only 7 % of the world's arable land, so land contamination and food safety have substantially threatened the food security of the nation.

Over the past decade, the Chinese central government and local governments were looking for countermeasures to remediate and prevent land contamination. A growing volume of literature has suggested nationwide contaminated land legislation in China. This article will argue that proper schemes to improve the soil/land environment should be incorporated into the contamination law at the national level. International experience should be carefully studied while developing a contaminated land regime in China. The structure of this article is as follows: first, it briefly reviews the scale, causes and impacts of land contamination in present-day China; then, it discusses the legal frameworks and existing legal schemes related to contaminated land management in China; finally, regimes that should be considered by Chinese environmental legislatures and policymakers are recommended based on a comparative study to world approaches.

3.2 The Growing Problem of Land Contamination in China

3.2.1 Status Quo of Land Contamination in China

Land contamination in China was, for a long time, perceived as involving relatively rare environmental pollution incidents. But today, a growing number of people are starting to realize that land contamination is a widespread problem as an inherent consequence of historical industrial and waste disposal practices. From 2005 to 2013, the MEP and the Ministry of Land Resources jointly conducted the first National Soil Pollution Survey, which covered 2.4 million square miles of land across mainland China (China MEP 2013). In April 2014, MEP released the results of the survey. As the report showed, approximately 16.1 % of the country's total lands, including 19.4 % of farmland, have been polluted. Human industrial and agricultural activities are considered to be the main sources of land pollution. More specifically, factory waste products, irrigation of land with polluted water, the improper use of fertilizers and pesticides and livestock breeding have all contributed to the farmland pollution (China MEP 2013). The survey also found that 82.8 % of the polluted land was contaminated by inorganic materials, such as cadmium, nickel and arsenic. Moreover, the levels of these materials in the soil had risen sharply since the 1990s. The level of cadmium had risen by 50 % in the southwest and in coastal areas and by 10–40 % in other regions. While geographical factors have been considered, the overall soil contamination situation in southern China is much worse than in northern China (China MEP 2014).

3.2.2 Causes of Land Contamination in China

Generally speaking, land contamination can be caused by both municipal and industrial pollution activities and farming practices. In the context of dramatic economic and social development over the past decades, the accumulative effect of the industrial and agricultural pollution is thought to be the main reason for the nationwide land contamination (China MEP 2014). Pollution from mining industries, the overuse of fertilizer and pesticides and polluted water irrigation on arable lands are the main contributors to land contamination.

In China, farming practice has been identified as one of the main causes of land contamination, with more and more people noticing the increasingly adverse effects (Grazer 2009). Typical actions include the over-application of pesticides, chemical fertilizers, animal and poultry excrement and sludge; sewage irrigation; discarded residue of agricultural plastic membranes; and mulching, most of which can be classified as nonpoint source pollution (NSP) (Ritter and Shiromohammadi 2001, Zhao 2013). At present, nearly one-third to one-half of all NSP arises from farming practices. As suggested by an investigation in 2009, about 100,000 km^2 of agricultural land has been polluted by pesticides, chemical fertilizer, wastewater irrigation, heavy metal pollution and plastic film (Zhao 2013). Among others, researchers have pointed out that over-application of chemical pesticides and wastewater irrigation are the two most important causes of agricultural land contamination in China (CCICED 2006; Zhu 2007).

As China is the world's main chemical pesticide producer and the world's largest chemical pesticide consumer, over-application of chemical pesticides has become one of the main sources of land contamination. In 2007, China exceeded the USA and became the world's largest chemical pesticide producer for the first time, with a total yield of about 173,000 tons (Zhao 2013). As the National Statistics Bureau of China (NSBC) revealed, there was a notable increase in the chemical pesticide yield from the year 2001–2010. According to the MEP, the total application of chemical pesticides in 1983 was 862 thousand tons; in 2006, pesticide consumption in China was over 1.31 million tons, and the average dosage deposited on agricultural land was 13.97 kg/ha, which is higher than that of developed countries (Conradie et al. 2000). By 2008, the application of chemical pesticides in China had risen to approximately 1.672 million tons (National Bureau of Statistics of China and Ministry of Environmental Protection of China 2009).

Low efficiency in the use of chemical pesticides is another problem. Although the total application of chemical pesticides in China is extremely high, the utility rate is as low as 30 %, which means approximately 70 % of the pesticides diffuse directly into the environment and accumulate in soils (Gould and Lewis 2009). In some high-yield districts, the frequency of spraying pesticides may be as high as 30 times per year, with the average amount exceeding 300 kg/ha, an amount which has been found to be increasing at a rate of 10 % per year (Zhu 2007). Meanwhile, a notable feature of chemical application in China is that chemicals with high toxicity account for a prodigious proportion of the whole range of chemical pesticides. Research has shown that more than 30 types of organ phosphorus pesticides with a total weight of 200,000 tons have been applied in China, among which more than 80 % were of severe toxicity (Zhu 2007). In addition, persistent organic pollutants (POPs), such as polychlorinated biphenyls (PCBs) and polynuclear aromatic hydrocarbons (PAHs), and compound pollution worsen the land contamination situation in the main corn-yield areas (Project Group 2007). During a 2007 study, POPs were widely detected in the topsoil of the Yangtze River Delta, and the average radium concentrations were higher than the traditional organochlorine pesticides (such as DDT), which was considered to be a serious new territorial soil contamination problem (Project Group 2007). Researchers therefore suggested that special attention be paid to farmlands within the Yangtze River Delta, which is thought to be the most significant region of cropland in China (Project Group 2007).

Large-scale wastewater irrigation was encouraged in northern China for years, especially in those areas suffering water shortages (Luo et al. 2009). Every year, China generated a huge amount of contaminated wastewater. Up to 1998, the area of farmland using wastewater irrigation was over 36,200 km^2, which accounted for 7.33 % of the overall irrigation areas and 10 % of the surface water irrigation areas.

That was 1.6 times larger than it was at the beginning of the 1980s (1977–1982) (Zhu 2007). By 2009, the amount of contaminated wastewater generated in China was 5.89 billion tons, of which industrial wastewater and municipal sewage sludge accounted for 2.43 billion tons and 3.55 billion tons, respectively (MEP China 2010a, b).

3.2.3 Impacts of Land Contamination

We already know that our agriculture-based life highly depends on the soil. Without soil, land plants cannot grow, and animals cannot survive. But without healthy soil, pollutants may enter into the food chain and cause harm to human beings and livestock. Land contamination, particularly the farming land contamination, is now threatening the quality of agricultural products, as well as the food security of the nation.

On the one hand, land contamination has greatly damaged taste and the quality of agricultural products and, in some cases, made them unsuitable for eating because of chemical residue. Hunan Province, a large rice-growing area in the middle of China, has some of the worst land contamination because it is one of China's top producers of non-ferrous metals. Hunan produced 16 % of the country's rice in 2012. However, in 2003, officials in Guangdong Province found that some rice had excessive levels of cadmium, and most of that rice was from Hunan Province (Wongapril 2014).

Land contamination not only raised concerns about domestic food quality but also worried overseas food importers like the USA. According to testimony before a subcommittee of the House Foreign Affairs Committee in May, food products of Chinese origin have dominated the USA food market in certain categories of food (Lin 2013). Chinese food imports are also expected to increase by about 10 % annually until 2020. Faced with the huge volumes of foods imported into the USA every year, the Food and Drug Administration was only able to inspect about 2.3 % of the total imports from China in 2011 (Lin 2013).

Tobacco, one of the most important economic crops in China, has also been polluted for the reason of land contamination. As some recent studies revealed, the quality of tobacco has been degraded. A comparative study of cigarettes produced in China and other countries in 2010 revealed that heavy metals in 13 Chinese cigarette brands exceeded the relevant standards; for some Chinese brands, the levels of lead, cadmium and arsenic were nearly three times higher than Canadian cigarette brands. This means that approximately 350 million smokers in China are potentially exposed to health risks resulting from land contamination (People's Daily Online 2010).

On the other hand, land contamination, together with salinization, desertification, erosion and radioactive pollution (Li 2006), has aggravated the shortage of arable lands and accordingly endangered the food security of China. China has 22 % of the world's population but no more than 7 % of the world's arable land.

Food security has been a chief mission and a primary state objective in the early twenty-first century. Environmental pollution, population growth and climate change are all factors that have worsened the food security of China. Scientists have indicated that around 36 million hectares of agricultural land have been contaminated by organics such as petrochemicals, pesticides and polycyclic aromatic hydrocarbons (Chen 2002). A number of traditional residential areas, like the Taihu Lake area, Yangtze Delta and the Pearl Delta have been suffering from serious pollution and eutrophicational problems over recent decades (Liu et al. 2005), and increasing discharges of heavy metal pollutants like lead, copper and zinc into the basin are thought to be main reasons (Jin et al. 2010).

3.3 The Current Legal Landscape for Regulating Contaminated Lands

3.3.1 Legislative Frameworks

The Chinese government has established a comprehensive environmental system of laws to address environmental protection issues. The regulatory framework comprises: (1) the Constitution of China, (2) Environmental Protection Law (amended 2014) (3) other nationwide laws and regulations, (4) decrees issued by the MEP (formerly SEPA), (5) local legislation, and (6) environmental protection standards.

Under this framework, air pollution, water pollution, solid waste pollution and noise pollution have been addressed by the constitution and environmental laws. Some have been amended several times to take into account dynamic socio-economic circumstances. Land contamination, however, did not draw much attention from the public before the 1980s (Qi 2009). Although there are no specific laws to deal with soil contamination in China, existing laws and regulations can be applied to soil contamination issues.

3.3.1.1 The Constitution of the People's Republic of China

The Constitution of the People's Republic of China (Adopted by the 5th National People's Congress, December 4, 1982) is the overarching law of China. A provision for environmental protection was first inserted in the Constitution in 1978. The Constitution states in Article 9:

> The State ensures the rational use of natural resources and protects rare animals and plants. The appropriation of or damages to natural resources by any organisation or individual by whatever means is prohibited.

In addition, Article 9 provides: 'the State protects and improves the living environment and the ecological environment and prevents and remedies pollution and other public hazards…', and thus '…organisations and individuals using land must ensure its rational use'.

The Constitution shares a basic value of protecting natural resources and controlling environmental pollution; more importantly, it emphasizes that environmental protection of the soil is a duty of the state.

3.3.1.2 Environmental Protection Law of the People's Republic of China (As Amended in 2014)

The Environmental Protection Law of the People's Republic of China (EPL) is the principal law for environmental protection in China (National People's Congress 2014). In April 2014, the EPL was amended and the new law will come into force starting 1 January 2015. Article 2 of the EPL defines 'land' as one element of the ecosystem. It has inherited the principle of coordinated development between economic construction, social progress and environmental protection and defines the rights and duties of both governments and individuals with regard to environmental protection. It emphasizes that the country should enhance the management of air, water and soil as well as establish relevant investigations, monitoring and access and remediation mechanisms. In addition, it also provides that the people's governments at all levels should provide for the protection of the agricultural environment by preventing and controlling land contamination, desertification and other negative impacts to land; by extending the scale of comprehensive prevention and control of plant diseases and insect pests; and by promoting a rational application of chemical fertilizers, pesticides and plant growth hormones. The central government should provide financial support by creating a special budget for the prevention and control of land contamination. The EPL also provides instructions for pollution prevention and abatement.

3.3.1.3 Other Nationwide Laws and Regulations

Other than the Constitution and the EPL, numerous provisions under laws and regulations at lower levels, which are embodied in laws, administrative regulations, bylaws and governmental documents, can also be applied to deal with land contamination problems. Areas covered by those provisions include pollution prevention, natural resource protection, city planning, agricultural environment protection and ecosystem protection. For example, according to the Law of the People's Republic of China on the Prevention and Control of Environmental Pollution by Solid Wastes (National People's Congress 2004a), local governments should bear the costs of dealing with discarded installations or sites that have historically been used for the purpose of the storage or treatment of industrial solid wastes (Article 35). This law can be applied when the liable party cannot be found, or the liable parties exist, but they have insufficient capacity to discharge their liabilities.

In addition, land contamination has been addressed by laws related to land resource and agriculture management. The Land Administration Law of the

People's Republic of China (National People's Congress 2004b) provides that people's governments at all levels should take measures to maintain and protect irrigation and drainage facilities as well as ameliorate the soil to increase fertility and prevent desertification, salinization, water loss, soil erosion and pollution (Article 35). In addition, the Agriculture Law of the People's Republic of China (National People's Congress 2002) incorporated regimes for regulating soil contamination by encouraging the rational use of chemical fertilizer and pesticides management to improve the quality of arable land. The law has introduced a registration system as a way of insuring integrated agricultural environment management (Article 58). Registration and/or licences are required with respect to the production of pesticides, veterinary medicines, fodder and feed additives, fertilizers, seeds and farm machines, which may endanger the safety of human beings and/or livestock (Article 21). Similar provisions on agricultural product safety can be found in the Regulation on Pollution-free Agricultural Product Management (2002),which requires that the environmental conditions of the areas producing pollution-free agricultural products should meet specific environmental quality standards (Article 9 (1)). Moreover, the Rules on Land Reclamation introduces a principle of 'those who destroyed the land should be responsible for reclamation work' to reclaim lands damaged by digging or sinking during the construction process (China State Council 1989).

Finally, laws on water protection and general ecosystem protection can also be applied to regulating land contamination. For example, the Rules for Implementation of the Law on the Prevention and Control of Water Pollution (China State Council 2000) provides that in cases where industrial wastewater or municipal sewage is used for irrigation, agricultural administration authorities of the people's governments at the county level and above should organize periodic monitoring of the quality of the irrigation water, soil and agricultural products and carry out appropriate action to prevent contamination of soil, groundwater and agricultural products (Article 24). Moreover, the Regulation of the Nature Reserves of the PRC (China State Council 1994) prescribes that productive installations should not be built in the core areas and buffer zones of nature reserves (Article 32(1)).

3.3.1.4 MEP Decrees and Environmental Standards

As the most important environmental protection authority of China, the MEP, the former State Environmental Protection Agency (SEPA), has promulgated numerous decrees with respect to land contamination problems. For example, the secretarial office of the former SEPA promulgated an administrative notice that deals with contaminated land remediation issues when a company changes its location (China SEPA 2004). The notice requires that all the industries and laboratories generating hazardous wastes and all the entities yielding or trading hazardous wastes should provide a soil function restoration plan based on a series of monitoring and testing and submit the monitoring report together with the restoration plan to the relevant environmental administrative departments before terminating their businesses or changing the original land use purpose. Environmental protection departments at various levels should attach importance to land contamination during the process of land redevelopment (China SEPA 2004).

A draft MEP Decree for contaminated site management, the Interim Measures for the Soil Environment Management of Contaminated Sites (China MEP 2009a), was issued by MEP on 15 December 2009. This decree is so far the only nontechnical environmental decree for the management of contaminated sites. It consists of six chapters with 29 Articles (China MEP 2009d). It defines a 'contaminated site' as land whose soil has been contaminated for the reason of (1) toxic and harmful materials, (2) abandoned materials or (3) mining (Article 2). It imposes liabilities on those who fail to report soil environmental conditions when contaminated site remediation has been completed (Article 27(3)). In addition, the polluter and/or owner who has land use rights may be liable for the land contamination. Therefore, the assignees of the land use rights may be liable for investigation and remediation expenses if they failed to carry out an environmental assessment before they accepted the problematic land (Article 8). Moreover, the local governments shall bear the site remediation costs if no liable parties can be reached. As the most important decree for defining the contaminated sites and liability issues, the deliberative draft of the measures was approved by the commission conference of the MEP on April 2011 (Zhou 2011).

To date, there are no comprehensive technical guidelines or standards on land contamination/soil pollution monitoring, contaminated land assessment and soil remediation techniques at a national level. Environmental standards are promulgated by the MEP and/or other administrative agencies to regulate specific, technical or compliance issues. Those standards may fall into one of five categories, which include environmental quality standards, pollutant discharge or emission standards, basic environmental criteria, criteria for samples and criteria for methodology. The environmental quality standards and pollutant discharge standards are compulsory standards; both standards can be divided into state standards and local standards (Article 3) (China MEP 2009b). Land environment protection is highly dependent on two essential standards: the Environmental Quality Standard for Soils 1995 (China SEPA 1996) and the Environmental Quality Risk Assessment Criteria for Soil at Manufacturing Facilities 1999 (China SEPA 1999). The first standard is mandatory, but it regulates agricultural land based on crop safety only. The second one is not mandatory and fails to address contamination issues for future land use. Today, some typical environmental standards about site management standards and technical guidelines are being drafted and released for trial use, which include (1) soil environment quality standards, for example, the Soil Environmental Quality Assessment and Remediation Standards for the Exhibition Site (provisional for Shanghai Expo) (HJ 350–2007); 2) testing standards, which include the Soil Environmental Monitoring Technical Specifications (HJ/T166–2004), Groundwater Monitoring Technical Specifications (HJ/T164–2004) and Water Environment Monitoring Specifications (SL219–98) issued by the Ministry of Water Resources; and (3) monitoring standards, which include the Groundwater Quality Standards

(GB/T14848–93), Soil Environmental Quality Standards (GB15618–1995) and Soil Environmental Quality of Industrial Enterprises (HJ/T25–1999).

3.3.1.5 Local Regulations

For most of the local governments, there is no specific local regulation or statute with respect to land contamination issues. However, there are a few exceptions. Generally, cities that have historically suffered from severe industrial pollution have established rules to regulate contaminated site monitoring or remediation. For example, The Measures of Soil contamination Prevention and Control was promulgated in Fujian Province September 2015. A similar regulation was promulgated in Hubei Province in February 2016. Zhejiang Province issued the Zhejiang Province Solid Waste Pollution Prevention Regulation in March 2006, the Beijing Environmental Protection Bureau (EPB) issued the Site Environmental Assessment Guidelines in January 2007 and the Shenyang Environmental Protection Bureau and Shenyang Municipal Planning Land Resources Bureau jointly issued the Management Measures of Contaminated Site Remediation and Control of Shenyang City (trial implementation) in 2007 (Shenyang Environmental Protection Bureau 2007). In addition, the Chongqing Municipal Government issued the Notice on Strengthening Industrial Contaminated Site Treatment and Remediation in 2008, which details requirements for meticulous conduct in contaminated site risk assessments (Chongqing City Council 2008).

To summarize, the following issues are prescribed subject to these local regulations: First is provisions on farmland pollution control, for example, control of farmland nonpoint pollution, ecological agricultural enhancement, farmland preservation and reclamation, water source preservation and wastewater irrigation. Second, provisions on soil environmental protection in mining areas also provide broad issues, for example, the ‘environmental impact assessment’ (EIA), pollution source control within water areas and clean production methodology. Third, solid waste pollution control, landfill gas monitoring, medical waste control and radioactive control for nuclear power stations have also been prescribed by these local regulations, rules and regulations.

Lacking statutory standards obstructs the effective management of contaminated sites, particularly in some urban areas. As a result, most recently, a series of standards and technical guidance for contaminated site management is being drafted at a national level by research institutes commissioned by the EMP. For example, the Guidelines for Contaminated Site Risk Assessment (China MEP 2009c), Guidelines for Soil Remediation of Contaminated Sites (China MEP 2009d) and Technical Guidelines for Environmental Monitoring of Sites (China MEP 2010b) have all been published for comments at the time this paper has been submitted.

3.3.2 Limitations of Current Laws and Regulations

As noted earlier, in present-day China, land contamination has become one of the most serious problems that has profoundly impacted the population's health and living conditions and arguably the sustainable development of Chinese society. It is likely that along with rapid industrial development, agricultural modernization and urban construction, a growing number of contaminated sites will be discovered in the near future. Although China's central government together with its affiliated ministries have been responsive to this emergent crisis, it should be noted that the existing legal system is far from sufficient and effective, considering its fragmented governance, existing legislative and policy framework, weak administrative organization and capacity and limited technical knowledge. At present, apart from several local regulations, there is no comprehensive national land contamination law in China.

The first major challenge is the lack of integrated legislation as a response to contaminated land problems. A comprehensive national legal framework with clear policy objectives and principles is a prerequisite at the central government level for dealing with the land contamination problems. The existing legal system for contaminated land in China lacks integration due to its lack of integrated legislation and effective technical approaches. Experience in the context of China and the international communities and an effective regulatory system for contaminated land require, inter alia, systematic integration of all aspects, including assessment, remediation, spatial planning, aftercare and monitoring.

Another challenge is that there is an extremely limited supply of experienced administrators and dedicated organizations responsible for contaminated land at both central and local levels. Insufficient technical support for the management and treatment of contaminated land commonly exists under the current legal and policy context. Land contamination prevention and control is a relatively new concern in China, and the necessary expertise and experience is still underdeveloped. In addition, it is clear that the existing legal system for regulating contaminated land is not operating optimally. For example, the Soil Quality Environmental Standard of China was developed about 20 years ago; it has been heavily criticized as being unable to tackle the current problems effectively (Wang 2008). Although a new standard proposal has been under public review for 5 years, the MEP has still not given a final approval of it (General Office of MEP 2009).

In addition, according to the present environmental law and policy, financial incentives for reclamation of contaminated land have not been fully developed by policymakers. This operates as a barrier to effectively dealing with contaminated lands. Funding mechanisms appear in some local regulations. However, there are no developed coordinated approaches that would enable legislation to be properly implemented.

Finally, there are no proper solutions for resolving conflicts between central and local government interests. Provisions on soil protection and land contamination prevention and control could easily encounter weak local protectionism in China.

Local protectionism means that local governments let their own local interests prevail over national concerns. The implementation of central directives becomes particularly challenging considering four main local governance weaknesses: (1) modest administrative capacity of local officials; (2) low accountability, as governmental structures are often not yet developed enough to effectively monitor and evaluate policies besides top-down mechanisms; (3) weak rule of law, as local courts have weak institutional status; and (4) limited citizen representation and participation (OECD 2009a).

3.4 Addressing Challenges by Referring Experience of Other Jurisdictions

3.4.1 An Overview of Global Approaches to Contaminated Land Management

Land contamination is a worldwide phenomenon that is faced by both developed countries and developing countries. In a handful of developed countries, national contaminated land law, which includes both national and/or federal legislation and state or provincial legislation, has been developed since the early 1970s (Brandon 2013). This trend may be attributed to several major incidents of large-scale land contamination cases. For example, in the USA, the Love Canal incident is widely credited as being the catalyst for federal legislation on liability for cleanup of toxic sites, leading to the Comprehensive Environmental Response, Compensation, and Liability Act (CERCLA) of 1980 (Switzer and Bulan 2002). In the Netherlands, a heavy soil contamination incident at Lekkerkerk in 1980 (de Roo 2003) finally led to the introduction of the Soil Clean-up (Interim) Act and non-statutory Soil Clean-up Guidelines in 1983 (Brandon 2013; de Roo 2003). Those major trigger events or a series of smaller incidents have provided sufficient political pressure to legislate on the issue in heavily industrialized or highly urbanized countries (e.g. the USA, Netherlands, Germany and Australia). Some other countries enacted specific contamination laws because they are more likely to feel pressure to redevelop urban areas due to land scarcity and a high population density (e.g. the UK and Japan).

To some degree, the broad experience of developed countries in terms of land contamination management can be learnt by developing countries, particularly for countries in which the awareness of land contamination was only raised recently, after the 1990s. In fact, several leading countries, such as the Netherlands and the USA, have influenced several others to select a particular model of land contamination legislation. For example, the liability regime under the CERCLA in the USA has been emulated in Taiwan, in Japan, and partly in the provinces of Canada (de Roo 2003, p. 124).

A comparative study shows that an integrated framework of contaminated land laws may comprise the following key elements (de Roo 2003, p. 12). Firstly, it should contain detailed procedures for the management of land contamination, including site identification, investigation, assessment, remediation planning, remediation work, monitoring and aftercare. Secondly, it should specify the role of public authorities, private professionals and responsible parties at each procedural stage. Thirdly, it should also stipulate an information management system (e.g. registration scheme or a database of contaminated sites) and secure people's right to get access to those information sources. Finally, the health and safety protection for site workers and neighbouring residents should be adequately addressed under the framework.

China now remains heavily reliant on its general environmental protection legislation to regulate land-contaminating activities, and this has to date proven insufficient to enable China properly to address the problem of the land contamination and food safety crisis. With the nature and extent of land contamination problems and their impact on food safety better understood in China, the Chinese government has realized that it is urgent to develop a nationwide contaminated land law based on global approaches and broader international experience (Cui and Wang 2013).

3.4.2 *Incorporate International Experiences into Chinese Legislation*

3.4.2.1 Contaminated Land Risk Assessment Scheme

In the context of land contamination, risk assessment refers to the qualitative and quantitative evaluation of the risk posed to human health and/or the environment by the actual or potential presence of soil pollutants. In most developed countries, a risk assessment would be undertaken once a decision has been made by the relevant authority, based on the initial site investigation, that a site is sufficiently contaminated to necessitate further assessment (McCaffrey et al. 2003). The aim of the risk assessment process is to determine whether the level of contamination present at a site poses a sufficient threat to human or environmental health to make remediation necessary or advisable (Landcare 2012). Assuming remedial action is recommended, the process would then lead to an evaluation of the most appropriate remediation options for the site in question.

In the UK, the risk assessment process is called 'options appraisal' (Brandon 2013, p. 324). With the introduction of the new regulatory regime for contaminated land in section 57 of the Environment Act 1995, the UK has endorsed a risk-based framework within which contaminated land can be identified, assessed and managed (Luo et al. 2009). This framework adopts a rationale and terminology from accepted international approaches to the regulation of environmental media, and it embodies the fundamental distinctions between toxicity, hazard and risk that exist throughout environmental science. Several terms like toxicity, hazard and risk are introduced under a risk assessment scheme. The USA has also developed a complex, contaminated site risk assessment procedure under the CERCLA.

It is now commonly accepted that a clearly defined risk assessment procedure is supposed to be an integral component of the contaminated land law of China. However, in the context of China, numerous challenges exist in this process. For example, the lack of technical expertise available to analyse the potentially contaminating activity or its associated risks, insufficient data about a site and financial gaps for conducting a thorough investigation sit in question (Brandon 2013, p. 326).

3.4.2.2 Contaminated Land Remediation Scheme

The benefits of remediation are mainly reflected in improvements in human and ecosystem health or the avoidance of future degradation in health or environmental quality. Using treatment technologies in contaminated land remediation is encouraged by agencies in many countries because they are perceived as adding environmental value compared with other approaches to remediation such as excavation and removal, containment or covering/revegetation. The ‘added’ environmental value is associated with the destruction, removal or transformation of contaminants into less toxic forms, providing a greater flexibility for site reuse (Boyd 1999).

In the UK, a contaminated land remediation mechanism was created by Part IIA of the Environmental Protection Act of 1990, which provides a statutory framework for the identification and remediation of contaminated land in the UK. The Act provides a legal definition of contaminated land and the procedures involved in its identification and remediation or cleanup. It prescribes the respective roles of the Environment Agency together with the local authorities in securing the remediation of contaminated land and defines the procedure for determining and apportioning liability for the cleanup. Finally, contaminated land remediation issues have also been provided by some other environmental acts and regulations, for example, the planning regime and the provisions of the Environmental Damage (Prevention and Remediation) Regulations 2009 of the UK.

While deciding whether or not specific land should be remediated, a cost-benefit method should be considered. Generally, the costs of contaminated land remediation include the direct costs and indirect costs. The former are comprised mostly of capital and labour expenditures, which are also referred to as the ‘costs of cleanup’. In the UK, the DEFRA Circular 01/2006 (Dept. for Food and Rural Affairs (UK) 2006) requires that such costs be taken into account when the authority is considering the cost of the proposed remediation package: all initial costs (including tax) of the remedial actions, including feasibility studies, design, specification, management and making goods; ongoing management and maintenance costs; and any ‘disruption costs’, such as depreciation in the value of the land or other loss or damage likely to result from carrying out the remediation (Dept. for Food and Rural

Affairs (UK) 2006). In addition to direct outlays, there are also indirect costs that typically include opportunity costs and institutional costs. Under the circumstance that the benefits surpass the potential costs, the environmental authorities should consider initiating the remediation procedures.

3.4.2.3 Contaminated Land Liability Scheme

Among others, the question of 'who should pay' is a fundamental issue of the statutory regimes under various jurisdictions. In the UK, according to Part IIA of the Environmental Protection Act of 1990, the primary liability falls upon those who have caused or knowingly permitted the polluting substances to be in, on or under the land ('Class A appropriate persons'). If no Class A persons can be found, then the present owners or occupiers, called Class B appropriate persons, will be liable in default. Both the local authority and the Environmental Agency have played an important role under the liability regime. The contaminated land liability regime of the UK covers a broad range of commercial entities such as companies, banks, insolvency practitioners, trustees and parent companies.

In the USA, the Comprehensive Environmental Response, Compensation, and Liability Act (CERCLA) of 1980 (Pub L No 96–510) defines the responsibility for the costs of remediation and who should be responsible for existing contamination. It contains a stringent liability scheme and creates a new cause of action; the statute establishes the nature of the new liability, the identity of liable parties and the triggers for liability, remedies and defences to liability. Under CERCLA, liability may be attached to a wide range of persons who are called potentially responsible parties (PRPs).

It would be more complicated if the liability scheme for contaminated land remediation was considered in China. On the one hand, like the USA and many European countries, China has adopted the polluter pays approach in its environmental laws and regulations. However, it is thought that this approach is not an effective way to address the contaminated land issues because it is often difficult to find the original polluters of a site as they have already closed down or gone bankrupt before the contamination hazard is unearthed (Luo et al. 2009). On the other hand, in China, a large proportion of historic land contamination cases were caused by state-owned enterprises, which means liability often falls to the government (Boyd 1999).

As a result, certain characteristics are fairly constant under the contaminated land liability scheme of China: strict liability, joint and several liability and government liability. That is, where two or more persons are appropriate persons in relation to any particular remediation action, the authority should make a decision whether any one person should be treated as not being an appropriate person associated with the remediation issue. If sites need remediation, but in whole or in part no appropriate person can be found, the enforcing authority will take over the remediation. The UK has adopted the concept of 'orphan linkage' to describe this situation where either no owner or occupier can be found for some land or where those who would otherwise be liable are exempted by one of the statutory provisions (Zhao 2013, p. 150). This experience could be referenced by China as a possible solution for cases in which no appropriate person can be found.

3.4.2.4 Environmental Liability Insurance Scheme

Requiring environmental remediation insurance has become a common practice in different jurisdictions (Fogelman 2005). Most recently, environmental liability insurance is encouraged by the newly amended EPL of China. Before that, it was simply addressed by a MEP decree, the Guiding Opinions on Environmental Pollution Liability Insurance (SEPA Document No. 189). One problem is that the current insurance laws and regulations are too obscure, and issues about how the existing environmental liability laws and regulations are to be applied in a land contamination case are yet to be clarified.

A number of obstacles have been observed for developing an environmental pollution liability insurance scheme in the area of contaminated land management in China. First of all, the market demand is unclear. In China, the demand for environmental insurance products will largely depend on the attitude of local governments (Swiss 2011). The degree to which local governments introduce and rigorously enforce environmental liability laws and regulations will substantially decide the future of this insurance. Secondly, the historical data for contaminated sites is missing. Today, databases related to land usage and contamination situations are extremely insufficient. This makes it quite difficult to estimate loss severity and frequency with conventional methods (Xiong et al. 2007; Zhang and Wang 2010). Thirdly, Chinese insurance companies lack underwriting experience. Finally, there are insufficient legal remedies in Chinese courts over environmental pollution cases. In some cases, although the court decided in favour of a plaintiff, it is still unable to stop the violation or obtain payment for compensation.

In order to establish a sustainable proper means to prevent and remediate contaminated lands, governments at all levels should encourage the development of a pollution insurance market and let the 'invisible hand' of the market work for the benefit of society. First of all, it should define clearly the financial risks associated with environmental liabilities arising from land contamination and estimate their predictability. Second, data collection and filing should be initiated by environmental protection agencies. Well-constructed information systems will benefit the establishment of an environmental insurance system. Third, stakeholders of interest should be identified. One essential condition for a successful insurance solution is that key groups of interest (insurers and potential insured) are able and willing to be involved in the process. Fourth, it is necessary to introduce a compulsory insurance scheme for contaminated land issues. While considering the introduction of compulsory environmental insurance for operators, the law and policy makers should carefully examine the maturity and competitiveness of the insurance market. Fifth, staff training should be enhanced to expand the knowledge and experience of environmental management officials and insurance enterprises concerning environmental insurance (OECD 2009b, p 192). Finally, it should be noted that environmental liability insurance is not the only tool for eliminating environmental crises. Therefore, it is necessary to develop various forms of financial incentives to ensure that polluters are able to meet their obligations.

3.4.2.5 Financial Assistance for Contaminated Land Remediation and Redevelopment

Economics are fundamental to the progression of contaminated site cleanup programmes. In fact, lack of adequate and affordable financing has been reported as the most significant barrier to the reuse of a contaminated site in almost all countries that face the contaminated land problem (OECD 2009b). Substantial expenditure and excess costs in terms of contaminated land reclamation are usually referred to as the 'gap' in financing (Davis and Sherman 2010, p. 151). It necessarily involves considerable expenditure in the form of site clearance, fees for consultants and analysts, removal or treatment of contaminated soil or groundwater and gas control or other remedial measures (Tromans and Turrall-Clarke 2008, p 495). The high costs of cleanup or remediation will inevitably act as a considerable barrier for investment in those areas needing reclamation works (Bartsch and Wells 2003). Therefore, a financial assistance scheme that can generate sufficient funds to cover the reclamation costs has been considered to be a necessary mechanism for contaminated land management.

In the UK, as prescribed in the Environment Act of 1995, s 37 (l) (b), the Environmental Agency has general statutory powers to carry out any engineering operations that are conducive to or incidental to the carrying out of its functions. In principle, the funding for work such as is necessary could come from government grants or borrowing or from government loans, including financial support from the Department for Environment, Food and Rural Affairs (Defra). Under the UK system, there are two major sources of contaminated land funding. The one is funding for expenditures made by authorities, especially local governments, to deal with the so-called orphan sites, where the original contaminator cannot be found and the authority becomes the owner or occupier. The other is from special funding mechanisms set up for the reclamation of contaminated land set up in recognition of the possible considerable expenditure of remediation (Zhao 2013, p. 223).

In the USA, financial assurance obligations are both voluntary and compulsory. It can be negotiated between EPA and PRPs as part of the superfund settlement process and be included in the terms of an enforcement order (Brandon 2013, p. 157). Financial assurance mechanisms include surety bonds, corporate guarantees, trust funds, insurance policies, corporate financial tests and letters of credit. Occasionally, other mechanisms may be used to suit specific circumstances, such as liens over real estate property, deposit accounts, escrow accounts and certificates of deposit.

Based on the above comparison, several key issues should be considered by the Chinese legislature to develop a proper funding mechanism for contaminated land remediation. Among others, fairness should be the primary concern and a significant objective of the funding mechanisms. This is because, compared with other values of the regime, such as the effectiveness or long-term sustainability, fair distribution of liabilities among all potential parties seems to be the issue causing the most concern in Chinese society. This will consequently require fair procedures for granting applications, such as criteria for ranking and screening applications for funding purposes, and a fair treatment to areas with different development levels. In addition, screening criteria for the application of financial instrument should also be well defined (US EPA 2010).

3.5 Concluding Remarks

This article analysed causes and potential impacts of land contamination in China. Among all possible contributors, land contamination is probably the most basic one that is continually threatening food safety in the long food-supply chain. If, 10 years ago, it was a contentious issue whether a specific contaminated land law should be established in China, it has now been commonly accepted that land contamination prevention and control measures should be taken within a timely manner to tackle the environmental crisis in China, among which food safety is one of the most urgent ones.

Over the past decades, environmental laws and regulations have been developed in China to address the soil environmental protection issue, and a number of fragmented mechanisms have therefore been established. However, that has proven to be far from enough to address all the challenges raised by land contamination and to tackle the food quality and safety problems facing present-day China. The historical legislative practices of China and outside jurisdictions have shown that a nationwide contaminated land law, including a systematic contaminated land remediation mechanism, a sounder liability scheme, a more effective financial assistance scheme and relevant schemes alike, should be developed to stop and control the prevailing land contamination all over the country.

This article aims to direct attention towards making a new law to improve environmental protection by maintaining good soil quality of arable lands and safeguarding the ecological system as a whole. However, this is only a theoretical assumption. A credible, comprehensive law on land contamination could provide relatively immediate benefits and substantial improvements to the soil environment if it were widely promoted by Chinese legislatures. Moreover, if China would draw on the combined experience of developed countries over the past decades and could adopt proper mechanisms based on its domestic conditions, the making of a nationwide contaminated land law would substantially mitigate the impacts of its land contamination crisis.

****

References

Chen, Huaiman (2005) Huan Jing Tu Rang Xue. Science Press, Beijing, p 8.

Wong, Ming-Hung and Bradshaw, Anthony D. (2002) The Restoration and Management of Derelict Land: Modern Approaches. World Scientific Publication, Singapore, p 19.

Wong, Edward, Pollution Rising (2013) Chinese Fear for Soil and Food. http://www.nytimes.com/2013/12/31/world/asia/good-earth-no-more-soil-pollution-plagues-chinese-countryside.html?_r=0. Accessed 20 May 2015.

China MEP (2013) Soil Pollution and Physical Health. Environmental Science Publishing House, Beijing.

China MEP (2014), Gazette of the National Soil Pollution Survey. http://www.zhb.gov.cn/gkml/hbb/qt/201404/W020140417558995804588.pdf. Accessed 20 May 2015.

Grazer, Walter (2009) Catholics going green: a small-group guide for learning and living environmental justice. Ave Maria Press, USA

Ritter, William F. and Shiromohammadi, Adel (eds) (2001) Agricultural nonpoint source pollution: watershed management and hydrology. CRC Press, USA, p 1.

Zhao, Xiaobo (2013) Developing an appropriate contaminated land regime in China: lessons learned from the US and UK. Springer, Heidelberg, p 11.

CCICED Special Policy Groupon Environmental Issues Facing New Rural Development (2006) Environmental Issues and Countermeasures Facing New Rural Development in China. http://www.cciced.net/encciced/policyresearch/report/201205/U020120529451221797895.pdf. Accessed 20 May, 2015.

Zhu, Wendong (2007) The Impact of Sewage Irrigation on Farm Production. Journal of Anhui Agricultural Science 35(12): 3623–3635

Conradie, Ernst M. et al (2000) A Rainbow over the Land: A South African Guide on the Church and Environmental Justice, Western Cape Provincial Council of Churches, South Africa.

National Bureau of Statistics of China and Ministry of Environmental Protection of China (2009) China Environment Statistical Yearbook. China Statistics Press, Beijing.

Gould, Kenneth A. and Lewis, Tammy L. (eds) (2009) Twenty Lessons in Environmental Sociology. Oxford University Press, London.

Project Group of National Program on the Development of Basic Research (973 Programme) (2007) Newsletter for the Soil and Air Quality Changing Regularity and Rule of Control in Yangtze River and Pearl, 2007.

Luo, Lei et al (2009) An inventory of trace element inputs to agricultural soils in China. Journal of Environmental Management 90(8): 2524–30. doi: 10.1016/j.jenvman.2009.01.011

Wongapril, Edward (2014) One-Fifth of China's Farmland Is Polluted, State Study Finds. Available via http://www.nytimes.com/2014/04/18/world/asia/one-fifth-of-chinas-farmland-is-polluted-state-report-finds.html?_r=0. Accessed 20 May 2015.

Lin, Ching-Fu (2013) Taking China's Food Safety Problem Seriously (I). https://blogs.law.harvard.edu/billofhealth/2013/10/10/taking-chinas-food-safety-problem-seriously-i/. Accessed 20 May 2015.

People's Daily Online (2010), Some Chinese cigarettes high in heavy metal content. Avail via http://english.peopledaily.com.cn/90001/90782/90872/7160898.html. Accessed 20 May 2015.

Li, Zijun (2006) Soil Quality Deteriorating in China, Threatening Public Health and Ecosystems. http://www.worldwatch.org/node/4419. Accessed 20 May 2015.

Chen, Huaiman (2002), Behavior of Chemicals in Soils and Environmental Quality. Scientific Publishing Company, Beijing.

Liu, Enfeng et al (2005) Geochemical features of heavy metals in core sediments of northwestern Taihu Lake, China. Chinese Journal of Geochemistry 24(1): 73–81

Jin, Zhangdong et al (2010) Concentrations and contamination trends of heavy metals in the sediment cores of Taihu Lake, East China, and their relationship with historical eutrophication. Chinese Journal of Geochemistry 29(1): 33–41.

Qi, Xiaomiao (2009) Contaminated Land Remediation in China is Still a Vacancy-an interview to Zhou Qixing. Science News Bi-weekly 20.
National People's Congress (2014) The Environmental Protection Law of the People's Republic of China. Order No 22.
National People's Congress (2004a) The Law of the People's Republic of China on the Prevention and Control of Environmental Pollution by Solid Wastes. Order No 31.
National People's Congress (2004b) The Land Administration Law of the People's Republic of China. Order No 28.
National People's Congress (2002) The Agriculture Law of the People's Republic of China. Order No 81.
General Administration of Quality Supervision, Inspection and Quarantine of the People's Republic of China (2002) Regulation on Pollution-free Agricultural Product Management (2002).
China State Council (1989) The Rules on Land Reclamation.
China State Council (2000) The Rules for Implementation of the Law on the Prevention and Control of Water Pollution. Order No 284.
China State Council (1994)The Regulations on Nature Reserves of the PRC. Decree No 167.
China SEPA (2004) The Notice on Readying for Preventing Environmental Pollution when Enterprises Make Move. Decree No 47, 1 Jun 2004.
China MEP (2009a) Interim Measures for the Soil Environment Management of Contaminated Sites (Draft). Decree No 1321.
China MEP (2009b) The Measures on the Management of Environmental Standards. Decree No 3.
China MEP (2009c) Guidelines for Contaminated Site Risk Assessment. Decree No [2009] 1020.
China MEP (2009d) Guidelines for Soil Remediation of Contaminated Sites. Decree No [2009] 1312.
China MEP (2010a) Gazette of China's Environmental Protection.
China MEP (2010b) Technical Guidelines for Environmental Monitoring of Sites. Decree No [2010] 99.
Zhou, Wenyin (2011) Interim Measures for the Soil Environment Management of Contaminated Sites Got Approved by the EPM Committee Conference. http://www.zhb.gov.cn/zhxx/hjyw/201103/t20110325_207637.htm. Accessed 20 May 2015.
China SEPA (1996) Environmental Quality Standard for Soils. GB15618–1995.
China SEPA (1999) The Environmental Quality Risk Assessment Criteria for Soil at Manufacturing Facilities. HJ/T 25–1999.
Shenyang Environmental Protection Bureau (2007) The Measures for Contaminated Sites Governance and Remediation of Shengyang City. Decree No. 87.
Chongqing City Council (2008) Notice on Strengthening Industrial Contaminated Sites Treatment and Remediation. Decree No 208.
Wang, Shuyi (2008) Some Considerations on Promulgating 'Soil Contamination Prevention and Control Law', Law Review of Wuhan University 2008(3): 73–78(in Chinese)
General Office of MEP (2009) Announcement about Amending Soil Environmental Quality Standard. Decree No [2009] 918.
OECD (2009a) OECD Rural Policy Review: China, OECD.
OECD (2009b) Ensuring Environmental Compliance: Trends and Good Practices. OECD.
Brandon, Elizabeth (2013) Global Approaches to Site Contamination Law. Springer, p 100.
Switzer, C.S., Bulan, L.A. (2002) CERCLA: Comprehensive Environmental Response, Compensation, and Liability Act (Superfund). American Bar Association, Chicago, p 3.
Gert de Roo (2003) Environmental planning in the Netherlands: too good to be true – from command-and-Control planning to shared governance, 168–169.
Cui, Qingxin and Wang, Sibei (2013) The 12th Standing Committee of the National People's Congress Published the Legislation Plan. http://www.npc.gov.cn/npc/zgrdzz/2013-12/12/content_1816288.htm. Accessed 20 May 2015.
McCaffrey, C. et al (2003) Guidance on Assessing the Risk Posed by Land Contamination and its Remediation on Archaeological Resource Management. Environment Agency, UK. http://

www.helm.org.uk/upload/pdf/Full_Guidance_on_Assessing_the_Risk_Posed_by_Land_Contamination_and_its_Remediation_on_Archaeological_Resource_Management.pdf.

Landcare Research (2012) Risk assessment for contaminated sites in New Zealand. Available at http://contamsites.landcareresearch.co.nz/risk_assessment_fundamentals.htm. Accessed 20 May 2015.

Luo, Qishi et al (2009) Risk-based management of contaminated land in the UK: Lessons for China? Journal of Environmental Management 90(2): 1123–1134.doi: 10.1016/j.jenvman.2008.05.001.

Boyd, James (1999) Environmental Remediation Law and Economies in Transition. http://www.rff.org/Documents/RFF-DP-99-21.pdf. Resources for the Future, Washington.

Dept. for Food and Rural Affairs (UK) (2006) Department for Environment Environmental Protection Act 1990: part 2A contaminated land. http://www.defra.gov.uk/environment/quality/land/contaminated/documents/circular01-2006.pdf. Accessed 20 May 2015.

Public Law No 96–510, 94 Stat 2767 (1980) (codified as amended at 42 USC§§9601–75 (2006)).

Fogelman, Valerie (2005) Environmental Liabilities and Insurance in England and the United States. Witherby, UK.

Swiss Re (2011) Environmental Pollution Liability Insurance in China: A Bumpy Road Leading to A Bright Prospect. http://www.swissre.com/clients/insurers/property_casualty/Environmental_pollution_liability_insurance_in_China.html. Accessed 20 May 2015.

Xiong, Ying et al (2007) Suppositions of China's Environment Pollution Liability Insurance. Modern Law Science 29(1): 24–28 (in Chinese).

Zhang, Yingbin and Wang, Xiaohui (2010) Study on the Environmental Liability Insurance Legal System in China. Journal of Beijing University of Aeronautics and Astronautics (Social Science edition) 23(4): 72–75.

Davis, Todd S. and Sherman, Scott A. (2010) Brownfields: a comprehensive guide to redeveloping contaminated property (3rd ed). American Bar Association, USA.

Tromans, Stephen and Turrall-Clarke, Robert (2008) Contaminated Land, 2nd edn. Sweet & Maxwell, London.

Bartsch, Charles and Wells, Barbara (2003) Financing Strategies for Brownfield Cleanup and Redevelopment. Northeast-Midwest Institute, Washington, DC. http://brownfield.cberdata.org/files/Bartsch%20and%20Wells,%20 %282003 %29.pdf. Accessed 20 May, 2015.

US EPA (2010) Proposal Guidelines for Brownfields Revolving Loan Fund Grants. http://www.ct.gov/dep/lib/dep/site_clean_up/brownfields/bfrlfgrantfy11guidelines.pdf. Accessed 20 May, 2015.

Chapter 4
Environmental Management System in China: The Target Achieving Process

Zhen Jin

Abstract This paper indicated how insufficient authorities given to China's Environmental Protection Bureaus and its lower status in governmental hierarchy caused leniency in the enforcement of environmental policies and hence the lower level of compliance in the past.

The Target Responsibility System and the management system for TRS process, such as Target Allocating Process and Target Achievement Process, have been important methods for China to overcome the problems in enforcing environmental policies.

Moreover, China has implemented and improved the system of public engagement with sufficient environmental information disclosure while expediting the processing of the complaints and petitions reported by citizens.

Keywords The Target Responsibility System • Air Quality Standard • Air Quality Implement Targets • 12th Five-Year Plan on Air Pollution Prevention and Control in Key Regions • Performance Examination Mechanism

Abbreviations

AQIT	Air Quality Implement Targets
CCCPC	Central Committee of the Communist Party of China
EIA	Environment impact assessment
EMC	Environmental Model City
EPB	Environmental Protection Bureau
FYP	The National Five-Year Plan
GDP	Gross domestic product
IPE	Institute for Public & Environmental Affairs
NRDC	The Natural Resources Defense Council
PEM	Performance Examination Mechanism

Z. Jin (✉)
China Research and Communication Center (CRCC), Japan Science and Technology Agency (JST), Tokyo, Japan
e-mail: kinnsinn8@gmail.com

H. Kitagawa (ed.), *Environmental Policy and Governance in China*,
DOI 10.1007/978-4-431-56490-4_4

PITI	Pollution Information Transparency Index
PPP	Purchasing power parity
RESCs	Regional environmental supervision centers
RRP	Regional Restriction Program
TRS	Target Responsibility System
WPT	Work Plan Targets

4.1 Introduction

As a developing country, China achieved rapid economic growth in the past decades. According to The World Bank's data, China's gross domestic product (GDP) based on purchasing power parity (PPP) overtook the United States and became the world's largest economy in 2014 (The World Bank 2015). The progress of rapid industrialization and urbanization played a vital role in their economic growth.

China, as the world's factory, developed the biggest energy-intensive industry with huge energy demand for fossil fuels. According to US Energy Information Administration, China became the biggest energy user in 2011 and largest net importer of petroleum in 2013 (US Energy Information Administration 2015). The share of urban population reached 54 % in 2014 from 26 % in 1990 (United Nations 2014), which was another factor to push up energy consumption, including electricity and gasoline in China.

At the same time, China started to face substantial environmental issues in water, soil, and air. Air pollution has increased adversely in China since 2008, especially in Beijing-Tianjin-Hebei, Yangtze River Delta, and Pearl River Delta. These areas, covering only 14 % of the country's land area, have been called key regions as they have shared nearly half (48 %) of their population, about 71 % of GDP, and 52 % of coal consumption in 2010. More importantly, the emission levels of air pollutants in these key regions have been 3.6 times higher than the national average (Ministry of Environmental Protection of the People's Republic of China (MEP) et al. 2012).

In 2013, over 70 major cities in North China, covering 1430 thousand km^2 area or 15 % of national territory in total, experienced heavy smog during the months of January and December (XinHua News 2013). Particulate matter smaller than 2.5 micrometers (PM2.5) has been the major component of air pollution in China, and coal burning has been the biggest contributor of PM2.5 pollution in those regions responsible for 50 % and 70 % of such pollution (TsingHua University 2014).

In 2012, China released a new Ambient Air Quality Standard (MEP 2012), with an aim to set limits, for the first time, on PM2.5 to protect human health, which would be implemented nationwide in 2016. However, 74 biggest cities, with the pressing needs to improve air quality, were required to implement the part of new standards (SO_2, NO_2, and PM10) from December 2012 and to implement the full standards (pulses PM2.5, O_3, and CO) from 1 January 2013 (MEP 2012).

From 2012 to 2013, China issued three regulatory action plans with Air Quality Implement Targets to reduce air pollutants including PM2.5.

To ensure the implementation of the target, the Target Responsibility System (hereinafter referred to as TRS) was introduced and later adopted not only as environmental policies but also as policies for other areas, such as climate change policies. Under the TRS, national targets were allocated to local governments and enterprises as mandatory targets with punitive measures for the personnel in charge of environmental protection.

In this paper, the following issues will be examined:

1. The institutional background of TRS
2. The target allocation and target achievement processes of TRS management system
3. Recent progresses in TRS
4. The legal and political authorization of TRS and its challenges

4.2 Target Responsibility System in Atmospheric Pollution Prevention Policies

4.2.1 Institutional Background of the TRS

According to the "2014 Report on the State of Environment in China", only 8 out of China's 74 biggest cities and 16 out of 161 cities (including 87 cities required to fully implement the new standards from 2014) passed the new standards in 2014 (MEP 2015). This was little better than the result of the last year, in which only 3 cities reached the new standards (MEP 2014), but the situation of air pollution in China was still severe.

In most cases, the main problem of China's environmental policy was the lack of effective management system for the compliance and enforcement of regulations. On 1 March 2015, Chen Jining, the new minister of MEP, pointed out at his first press conference that "In the past, environmental laws have been enforced leniently, and the overlooking of violation has become normal" (China Economy Net News 2015). The low level of compliance was a good evidence to indicate what used to happen there. As the studies of the enforcement problems in Environmental Impact Assessment (EIA) Policy suggested, nearly 40 % of the staffs at the Environmental Protection Bureau (EPB) of the local governments believed that over 40 % of companies started construction without EIA permission. In some areas such as Sichuan Province, the violation rate was at least 50 % and some areas reached nearly 100 % in 2007 (Wang 2011).

The low rate of pollution charge collection was another evidence of the problems. In 2006, a close study on environmental compliance and enforcement in China revealed that average rate of pollution charge collection was less than 50 % of the charges imposed, which was certainly a factor to lessen incentives for compliance (OECD 2006).

Moreover, the EPBs were not formally granted sufficient enforcement authorities to order the state-owned enterprises under the old China's environmental law until the end of 2014, even if such enterprises violated the regulations (Ma and Ortolano 2008). Similarly, the law did not give any authorities to county-level EPBs to order any enterprises directly under the jurisdiction of the central government or the provincial governments to comply with the regulations, even if such enterprises violated them (Article 29).

From 1 January 2015, however, the enforcement of new environmental law has allowed the EPBs at or above the county level to issue orders to all types of enterprises to restrict production or emissions or to suspend production for rectification (Article 25 of New Law).

In spite of the improvement as mentioned above, there still remains a problem that the EPBs have lower administrative status in the governmental hierarchy. Because the EPBs come under those local governments institutionally and financially, their enforcement actions are easily influenced by local government leaders who have greater interests in promoting GDP growth than protecting environment (OECD 2006). TRS is one of the important methods to address such problems in China. The following three points indicate how TRS works:

- TRS is a legal scheme with an aim to clarify the responsibility of local governments in attaining the concrete legally binding targets allocated by the central government through National Target Allocating Processes.
- Under the TRS, the provincial-level governments are required to achieve the allocated targets and shall take joint liability with the lower levels of local governments under their direct control.
- In order to ensure the effectiveness of TRS, the law is set in a way that the achievement of environmental targets not only affects the personnel evaluation of local government officials in charge but also results in punitive measures imposed.

4.2.2 The Scheme of the TRS

The TRS was originally established through the State Council Document No. 29 in 2006 to allocate nine mandatory national targets set in the 11th national Five-Year Plan(11th FYP) to all local governments. The national targets were first translated into provincial targets and then further subdivided and redistributed by provincial governments to their respective lower-level governments. The central government placed particular emphasis on energy-saving and reducing emissions of major pollutants and combined these two policies into a single policy in 2007 by issuing the State Council Document No. 15 on 23 May 2007. During the period of 11th FYP, China did not set Air Quality Improvement Targets (AQITs) and only had Amount Reduction Targets for major air pollutants including SO_2 and NO_2.

4.2.2.1 Time Schedule of Target Setting Process

From 2012 to 2013, China issued three regulatory action plans with legally binding targets to control air pollution, especially to reduce PM2.5. On 5 December 2012, China released the "12th Five-Year Plan on Air Pollution Prevention and Control in Key Regions" (12th FYP action plan) to set a road map for air pollution control for the next 5 years. It was the first time the central government of China issued a comprehensive action plan with concrete air quality targets, which had legally binding force for local governments, covering 19 provincial-level jurisdictions with 117 cities (MEP et al. 2012). The plan required these key regions to reduce ambient concentration of SO_2 and PM10 by 10 %, NO_2 by 7 %, and PM2.5 (expected target) by 5 % by 2015 compared to 2010 level.

The plan also introduced more strict PM2.5 standards in three key regions, which would cover areas of Beijing-Tianjin-Hebei, Yangtze River Delta, and Pearl River Delta. In these regions, 47 big cities with the levels greater or equivalent to prefectures would be required to reduce 6 % of PM2.5 concentration by 2015 as a binding target. Subsequently, these PM2.5 targets were replaced by new targets as an emergency countermeasure for heavy smog started in early 2013.

On 10 September 2013, the State Council issued the "Atmospheric Pollution Prevention Action Plan" (New Action Plan) in response to huge social unrest caused by severe air pollution in the same year (The State Council of the People's Republic China 2013a). The New Action Plan stipulated that by 2017 (compared to 2012), it would be required:

- To reduce the urban concentration of particulate matters (PM10) by 10 % in all cities greater than or equal to prefecture level and to gradually increase the annual number of days with fairly good air quality
- To reduce PM2.5 concentration levels by 25 % in Beijing-Tianjin-Hebei region, 20 % in Yangtze River Delta area, and 15 % in Pearl River Delta region
- To control the annual average concentration of PM2.5 in Beijing to below 60 micrograms per cubic meter

When compared with the 12th FYP action plan, the New Action Plan clearly strengthened the regulatory levels in three key regions.

After the issuance of "Implementation Rules of Action Plan Against Air Pollution in Beijing, Tianjin, Hebei and Surrounding Areas" (the Implementation Rule) by MEP and other four central government offices in September 2013, the provinces of Shandong, Shanxi, and Inner Mongolia were added as additional regulated regions, imposing the PM2.5 targets to reduce concentration levels by 20 %, 20 %, and 10 %, respectively, by 2017 (MEP et al. 2013).

By the end of 2013, 31 provincial-level regions were allocated Air Quality Improvement Target (AQIT) of PM2.5 or PM10. Among them, 11 regions were required to achieve AQIT of PM2.5, and another 17 regions imposed the AQIT of PM10 (Table 4.1). All targets should be achieved by the end of 2017 compared to 2013 (Table 4.1).

Table 4.1 Air Quality Improvement Targets in 31 provincial-level regions

Air Quality Improvement Targets (by 2017 from 2013)		Regions	
Annual average reduction of PM2.5 (%)	−25 %	Beijing, Tianjin, Hebei	11
	−20 %	Shanxi, Shandong, Shanghai, Jiangsu, Zhejiang	
	−15 %	Guangdong, Chongqing	
	−10 %	Inner Mongolia	
Annual average reduction of PM10 (%)	−15 %	Henan, Shaanxi, Qinghai, Xinjiang	17
	−12 %	Gansu, Hubei	
	−10 %	Sichuan, Liaoning, Jilin, Hunan, Anhui, Ningxia	
	−5 %	Guangxi, Fujian, Jiangxi, Guizhou, Heilongjiang	
	Continuous improvement	Hainan, Tibet, Yunnan	3

Source: MEP (2013a): The MEP has signed target responsibility agreements for preventing and combating air pollution with 31 provinces http://www.gov.cn/gzdt/2014-01/07/content_2561650.htm

The AQIT was decided through two different decision-making processes. Firstly, the AQIT of PM2.5 was decided by the State Council in top-down decision-making process through the announcement of New Action Plan. Secondly, the AQIT of PM10 was decided in bottom-up decision process through the Scheme of Responsibilities Document, in which each provincial-level government was demanded to submit "air pollution prevention and control responsibility documents" that contained AQIT with comprehensive work plans (MEP 2013a). The work plans were evaluated and approved by MEP if they satisfied the conditions of validity and feasibility between AQIT and the work plan.

4.2.2.2 Target Allocating Processes

In accordance with allocated AQIT, each provincial-level government was required to break down the target to sub-targets and reallocate them to its city- and county-level governments (The State Council of the People's Republic China 2013b). Both allocated targets and sub-targets would be subdivided into annual targets during the period from 2014 to 2017, and the achievement of annual targets would be evaluated in every other year (Table 4.2).

To ensure the implementation of allocated targets, the provincial-level governments were required to formulate detailed implementation rules and comprehensive work plans. The work plans were to contain the concrete amount of targets for ten key tasks: adjustment and optimization of industrial structure, environment-friendly production, coal management and vehicle fuel supply, rectification of small-size coal-fired boilers, air pollution control for industrial sector, dust control

Table 4.2 Rate of achievement for annual targets

PM2.5 Air Quality Improvement Target (by 2017 compared to 2013)				
Annual targets (rate of achievement)	Annual target (rate of achievement)			
	2014	2015	2016	2017
	10 % of all	35 % of all	65 % of all	100 %

Source: MEP (2014a): Detailed Rule for Performance Assessment Measures for Air Pollution Prevention and Control Action Plan (in Chinese) http://www.mep.gov.cn/gkml/hbb/bwj/201407/t20140725_280516.htm

in urban area, vehicle emission prevention and control, energy efficiency in buildings and heating, capital investment in the prevention and control of air pollution, and management of atmospheric environment (The State Council of the People's Republic China 2014).

The work plan should be divided into annual plans and then broken down to the lower-level governments and enterprises. Annual work plans would be required to contain the concrete amount of targets for ten key tasks, which would be approved by provincial-level governments.

4.2.2.3 Assessment Indicators for Target Achieving Performance

Both annual achievement of AQIT and annual Work Plan Targets (WPT) will be verified and reported every year, and the assessment results will provide significant basis for personnel evaluation of leaders and leading cadres in regions (The State Council of the People's Republic China 2014). Failure to achieve mandated targets will not only affect personnel evaluation of the local government officials in charge but also result in punitive measures.

In 2014, MEP issued the regulatory document "Detailed Rules of Performance Assessment Measures for Air Pollution Prevention and Control Action Plan", with an aim to clarify the formula for assessing the achievement of annual AQIT and WPT targets.

According to the document, the scoring method has 100 points as a full mark and calculates points depending on the ratio of target achievement for AQIT and WPT.

Table 4.3 shows the scoring method for assessing the achievement of AQIT. The full achievement of AQIT only counts for 60 points, and the only way to get another 40 points will be to have the local government overachieving the annual target at or above 30 %. In the case of failing to achieve the target, the score will be calculated by two different ways. If there are some improvements on the annual average air quality in comparison with the year before, the final score of annual target achievement will be calculated by the proportion of actual achievement over the annual AQIT target. In the case of no improvement and the worsening of the annual average air quality, it will be calculated as zero point. There is some rule applied

Table 4.3 Score for AQIT

Total score: 100				
Score for achievement of annual target: 60			Score for overachievement: 40	
Achieved	Non-achievement with little improvement on the annual average air quality comparison with the year before	Non-improvement and the annual average air quality worsen in comparison with the year before	Overachieved at or above 30 %	Overachieved less than 30 %
60	60 × (actually achieved target %/ annual improvement target %)	0	40	40 × (overachieved rate %/30 %)

to the score for over achievement. If the local government's overachievement rate is at or above 30 % in comparison with their annual target, it will get another 40 points fully. If it is less than 30 %, the score will be calculated by the proportion of actual overachievement rate over the required overachieved rate of 30 %.

Table 4.4 summarizes the scoring method for reviewing the achievement performance of WPT, which contains 10 indicators with 29 sub-indicators. Each sub-indicator has more detailed branch targets with the corresponding scores. All performance and efforts conducted by provincial-level governments with an aim to improve the air quality in their administrative districts will be reviewed by this scoring method.

4.2.2.4 Punitive Measures

The central government of China will use several punitive measures if a local government fails to achieve annual targets.

Firstly, as a punitive measure applicable to the region where a local government fails to pass the annual assessment, MEP will suspend EIA approval of projects under the program called Regional Restriction Program (RRP), so to halt all new construction projects, which will have deteriorating effects on regional air quality (with the exception of projects aimed to improve people's livelihood and/or to save energy). In addition, local governments are allowed to use the power of RRP over their administrative regions or branches of local governments if such regions or branches fail to pass the annual assessment. Moreover, RRP will not only be used in the case of failure to meet total emission caps and environmental quality targets but also in many other cases, such as the violation of EIA rules, and large-scale environmental accidents (Wang and Hao 2015). Since the first RRP imposed by MEP in 2006, there have been 120 cases of RRP imposed by the MEP and other environmental agencies of local governments by the end of July 2015 (Zhu et al. 2014).

Table 4.4 Scoring method for WPT

No.	Indicator	Score of indicator	Sub-indicator	Score of sub-indicator
Total score 100				
1	Optimize the industrial structure	12	New capacity control for overcapacity sectors	2
			Rectification of the illegal construction project in overcapacity sectors	2
			Elimination of backward production capacity	6
			Relocation of highly polluting factories	2
2	Environment-friendly production	6	Assessment of environment-friendly production and upgrading of the related technology for key sectors	6
3	Coal management and vehicle fuel supply	10	Coal consumption cap	0 or 6[a] or 8[b]
			Coal cleaning treatment	4 or 0[a, b]
			Clean coal supply for residential use	0 or 2[a]
			Diesel and gasoline fuel supply under China IV and V standards	6 or 2[a, b]
4	Rectification of coal-fired boilers in small size	10	Elimination of small coal-fired boiler	8
			Permit regulation for new coal-fired boiler	2
5	Air pollution control for industrial sector	15	Industrial dust (PM) control	8
			Industrial VOC control	7
6	Urban dust control	8	Dust control from construction site	4
			Permit regulation for new coal-fired boiler	4
7	Vehicle emission prevention and control	12	Out of use for yellow-label vehicles	7
			Environmental labeling management for vehicles	2 or 1[a, b]
			Promote the use of new vehicles	0 or 1[a, b]
			Capacity building for environmental supervision of vehicles	1
			Building transportation system for pedestrians and cyclists	2
8	Advancing the construction of energy-saving building with heat measurement	5	Advancing the construction of energy-saving building	5 or 2[c]
			Advancing the heat measurement	0 or 3[c]
9	Investment for air pollution prevention and control	6	Investments in air pollution prevention and control measures by	6

(continued)

Table 4.4 (continued)

Total score 100				
No.	Indicator	Score of indicator	Sub-indicator	Score of sub-indicator
			local governments, enterprises, and society	
10	Environmental management for air pollution	16	Annual work plan formulation	2
			Advancement in ledger management scheme	1
			Quality management and monitoring for atmospheric environment	5
			Agricultural waste control	3
			Controlling the agricultural waste incineration	1
			Environmental information disclosure	4

[a]Score for Beijing Municipality, Tianjin Municipality, and Hebei Province
[b]Score for Shandong Province, Shanghai Municipality, Jiangsu Province, Zhejiang Province, and Guangdong Province
[c]Score for the North China heating area, including Beijing Municipality, Tianjin Municipality, Hebei Province, Shanxi Province, Inner Mongolia Autonomous Region, Liaoning Province, Jilin Province, Heilongjiang Province, Shandong Province, Henan Province, Shaanxi Province, Gansu Province, Qinghai Province, Ningxia Hui Autonomous Region, and Urumqi Autonomous Region

As the second punitive measure against a local government failing the annual assessment, MEP will provoke the government's certification called Environmental Model City (EMC), which MEP itself has previously awarded to such local government. Because EMC is considered as an important certification requirement for getting environment subsidies, failure to meet the certification of EMC will make it harder for such government to get subsidies from the central government and/or other local authorities of higher level.

As the third punitive measure, the State Council or MEP will arrange interviews with a person who is mainly responsible for the poor air quality management of cities or regions. If necessary, the State Council will arrange talks with the leaders of provincial-level government as well. MEP will also implement interview directly or indirectly with different levels of local governments based on the regulation document called "The Provisional Rules for EMP to arrange the talks" issued in 2014 (MEP 2014a). Indirect interviews will be conducted by regional environmental supervision centers (RESCs), which cover several provincial-level governments, as the branch office of EMP.

On 15 September 2014, the RESC of Hua Nan arranged the first "talks" with the mayor of Hengyang City for the negligence of sewage treatment measures. Until August 2015, EMP and RESC made 22 interviews with cities (Time Weekly 2015).

In the talks, the mayors have been asked to attach greater attention to air quality improvement, to introduce effective plan, and to set concrete targets, in return for the promise of allowing them to continue their political career. The talks exert great pressures on the mayors, because all the processes of the talks have to be opened to the public. After being interviewed, the mayors are to take prompt measures such as to shut down the factory that has violated environmental laws, orders, and rules and to impose disciplinary actions with warnings, dismissals, and criticisms to punish the person in charge of environment protection policies.

Finally, as the fourth punitive measure, the failure to pass assessment may lead to the lower performance evaluation of individuals, especially of those individuals at or above the managerial level. As the achievement of TRS will be directly linked with the evaluation of individual performance, it has given essential motivation for the leaders of the local governments to make the best efforts for achieving the allocated targets. If they fail to fulfill the targets, the person who is in charge of such region or local government will receive negative assessment in his/her personnel evaluation and may invite warnings, dismissal, or criticism.

4.2.3 Performance Examination Mechanism

The common understanding about China's Performance Examination Mechanism (PEM) is that the "causes" set forth by the (Chinese Communist) Party and (Chinese State) Government will play the central and fundamental roles not only to ensure the effectiveness of policy implementation but also to strengthen the political control power over the country.

Many studies suggest that PEM should be improved. For a long time, the regional economic development used to be the most important index to evaluate the performance of causes. That is the main reason why some leaders from the local governments tend to take indifferent attitudes toward environmental protection policies.

In November 2013, in order to modify the way which focusses only on GDP achievement as the major factor of PEM, the third plenary session of the 18th Party Congress issued "The Decision of the CCCPC on Some Major Issues Concerning Comprehensively Deepening the Reform", to indicate that GDP would no longer be taken as the measurement of PEM in areas of restricted development and for the ecologically vulnerable key counties that account for more than half of the county-level governments in China (CIConsulting Net 2015).

In December 2013, the Organization Department of the CPC Central Committee released the Notification on Improving the Government Performance Examination among Local Party and Government Leader Group and Leading Cadres, which stipulated a new way to examine government performance by raising the weight of indicators such as environmental protection and safe production.

On 17 Aug 2015, the Central Committee of CPC together with the General Office of the State Council issued an additional document called the "Accountability Methods on Environment Damage Caused by Leaders and Cadres of the Party and Government (Trial)". The methods established therein set the punishment rules for officials who would cause serious damages to ecological environment and resources, even if they would be newly appointed or promoted to the positions or even if they have already retired (The Central Committee of CPC and the General Office of the State Council 2015).

In July 2014, Shanxi provincial government released the revised rates of PEM, which canceled the economic growth rate as PEM indicator, letting 36 county-level local governments relieved of GDP examination. After Shanxi, six more provincial-level governments, such as Fujian, Ningxia, Hebei, Zhejiang, Guizhou, and Shaanxi, have either canceled or revised the proportion of GDP examination (Li 2014).

4.2.4 Recent Improvement in Public Engagement

Fundamentally, the enforcement of regulations and legally binding targets relies on the appropriate verification process, for which it is essential to have sufficient public engagement and scientific Performance Examination Mechanism.

Chinese government has achieved good results from the timely establishment of environmental monitoring system nationwide, including air quality observation network. On the other hand, MEP obviously has difficulty in preventing every kind of corruption to seep into the system, such as the falsification of monitoring data by the administrative authority. In the paper released by XinHua, the vice minister of MEP, Wu Xiaoqing, points out that some local governments have forced the agencies in charge of air quality monitoring to falsify data, which has severely damaged the credibility of the government (XinHua Net News 2015).

It has been widely argued that China should further improve and implement the system of public engagement, enhance environmental information disclosure system, and introduce efficient processing scheme to expedite the processing of complaints and petitions reported by citizens.

In response, the Chinese government has started to give more attention to environmental information disclosure and has issued four related policy documents in 2013 and 2014 as follows (China Council for International Cooperation on Environment and Development (CCICED) 2014):

- On 15 October 2013, the General Office of the State Council released the document, "On Further Enhancing the Government's Ability in Information Disclosure, Responding to Public Concerns and Promoting Government Credibility."
- On 14 November 2013, the MEP issued the "Guideline of Government Information Disclosure on Environmental Impact Assessment of Construction Projects."
- On 31 March 2014, the MEP released the policy instrument, "On Further Enhancing the Work of Environmental Information Disclosure and Public Opinion Guidance."

- On 22 May 2014, the MEP released the "Instructions on Promoting Public Engagement in Environmental Protection."

A close study on the progress of environmental information disclosure in China was made at the Institute for Public & Environmental Affairs (IPE) and the Natural Resources Defense Council (NRDC), which showed, in full detail, the level of information transparency for 120 big cities (IPE and NRDC 2015). Since 2009, IPE and NRDC have released six annual assessment reports and used the Pollution Information Transparency Index (PITI) as core evaluation criteria. Table 4.5 shows the contents of PITI.

The latest report for 2014–2015 pointed out that the PITI average scores for 120 cities reached 44.3 point, representing 54.4 % increase from the previous year's score of 28.5 point.

One typical example of progresses in environmental information disclosure would be the disclosure of online monitoring information from 2013. In 2004, China started to establish its nationwide environmental monitoring network with the database that contained the real-time measured data, in order to improve the effectiveness of environmental management. For some years since then, however, the database had been available only for the internal use of MEP and EPBs, with the information inaccessible to ordinary citizens. Moreover, the protective attitudes taken by local governments made it more difficult for the timely disclosure and/or publication of high-quality data and information (IPE and NRDC 2015).

On 31 July 2013, MEP issued regulatory document called the Methods for Key State-Monitored Enterprise Pollution-Source Self-Monitoring and Information Disclosure (Trial), which required the key state-monitored enterprises to disclose the data of pollutant emissions, including real-time monitored figures, on website platform established by provincial-level governments (MEP 2013b). By the end

Table 4.5 Contents of the Pollution Information Transparency Index

Total score 100			
Indicator	Score of indicator	Sub-indicator	Score of sub-indicator
Environmental supervisory information	50	Publication of excessive emissions and other violation records	23
		Enterprise environmental performance assessment	5
		Publication of emission fees	2
		Disclosure of online monitoring information	20
Interaction and response	15	Petitions and complaints	7
		Disclosure upon application	8
Enterprise emission data	20	Disclosure of key enterprise emission statistics	16
		Publication of clean production audit information	4
EIA information			15

of 2014, 29 out of 31 provincial-level web platforms were launched, with the number of monitored enterprises increased to 14,920 (MEP 2014b). The web platforms have provided real-time monitored data of key monitored enterprises every hour for wastewater and exhaust gas emissions. However, few platforms could provide the data with the full coverage of enterprises, sufficient data integrity, higher rate of acquisition, and lower rate of deficiencies. More importantly, 19 out of 31 platforms have not been able to disclose real-time monitored data for wastewater and exhaust gas emissions (IPE and NRDC 2015).

One progress could be seen in the disclosure of EIA information. More than 30 years ago, China adopted EIA as an important measure to prevent environmental damages through the review of a project plan before the start of construction. Until today, China is still facing the problems of illegal constructions without permission from EIA. The reason why the EIA failed to play the important role originally expected by the government was that "China only copied the technical assessment aspect of EIAs from the West and ignored public participation, which was essential for a high-quality and effective EIA" (IPE and NRDC 2015).

Until late 2013, citizens were only allowed to access the abridged version of EIA documents, which was the main cause of discouraging public participation. On 14 November 2013, the MEP issued "The Guide for Governmental Information Disclosure Concerning EIA of Construction Projects (Trial)" which stipulates that the full text of all EIA reports should be disclosed from 1 January 2014. By the end of August 2015, 91 out of 120 cities disclosed the full text of EIA reports up from 42 cities in the year before (IPE and NRDC 2015). On the other hand, there would be a room to improve public participation procedures, still. According to the regulatory document mentioned above, the public comment periods of EIA should be 10 days. Due to the short notification period, however, citizens have been deprived of opportunities to make public comments as the comment period would likely expire by the time they have become aware of the release of EIA report.

Disclosing the resolution results of petitions and complaints is another area China needs to make progress. In recent years, the channels for making environmental complaints in China (reporting system) have been diversified. Citizens have a choice to make complaints not only through existing channels, such as telephone hotlines 12369 and web site to contact MEP directory, but also through new channels using social media like Weibo, as the communication channel between government and society. Since 2014, there are 70 cities using Weibo as the communication channel to receive reports of environmental violation and to disclose the information, including the subjects of petitions and complaints, the names of the accused with related statistical data, the status of investigation, the result of resolutions, etc. (IPE and NRDC 2015).

In addition, EPA has published monthly reports to disclose all cases of resolution results in order to ensure the effectiveness, integrity, and transparency of reporting system (EPA 2015).

4.3 Legal Analysis of TRS

4.3.1 Legal Background of TRS

Based on legal analysis, three main reasons for adopting the TRS method to ensure policy implementation include (1) strong command and supervisory authority of the central government over provincial governments, (2) strong binding force of FYP, and (3) greater power of the Communist Party of China (CPC) to control personnel affairs (Jin et al. 2013).

4.3.1.1 Strong Command and Supervisory Authority of the Central Government over Provincial Governments

There is no regional or local autonomy law in China. Hence, there is no legislation that defines the division of administrative responsibilities between central and local governments or the roles of central government in local matters. Furthermore, provincial governments are positioned as administrative bodies under the unified command and supervision of the State Council, and a clear command and supervision relationship is established between the central and regional governments (Constitution of the People's Republic of China and Article 55 of the Organization Law for Local People's Congresses and Local People's Governments) (Jin et al. 2013).

4.3.1.2 Strong Binding Force of the National FYP

FYP and its sub-plans are legally and strongly binding for local governments, which have the obligations to comply with the targets allocated to them by the State Council. China's Constitution gives the State Council the authority to draw up the FYP for national economic and social development and to implement it (Article 89, Clause 5). To implement the FYPs, the State Council has the authority to adopt administrative measures, enact administrative rules and regulations, and issue decisions and orders (Constitution, Article 89, Clause 1; Legislative Law, Article 56, Clause 2). The local governments are responsible for implementing the National FYP (Article 59, Clause 5 of the Organization Law). Moreover, the higher branches of government can reprimand local government heads for refusing to carry out their orders (Ordinance on Dealing with Administrative Body Public Servants [State Council Order], Article 19, Clauses 34–36: reprimands, warnings, demotions, suspensions, and dismissals) (Jin et al. 2013).

4.3.1.3 Greater Power of the Communist Party of China to Control Personnel Affairs

Personnel affairs of public servants in China, in particular those individuals at the managerial level (e.g., province heads, mayors, bureau directors, section heads, etc.), are uniformly controlled by the CPC. In other words, the authority to punish (except for removal from office) administrative heads lies with the higher branches of government, but the CPC also has de facto authority to appoint, promote, and transfer personnel to other posts (e.g., from an administrative body to a legislative body) after completing the term of office. For example, provincial heads are supposedly elected by Regional People's Congresses, but the authority to recommend individuals for such posts lies with the provincial Party Committees and the approval of senior officials (members of the Central Committee of the CPC) will be required (see Rules for Selecting Senior Party Officials, Article 8 and subsequent articles). The same process applies to the selection of central government leaders when there is a leadership changeover.

4.3.2 Legal Issues of TRS

4.3.2.1 Legally Binding Force of Allocated Targets

As mentioned above, allocated targets are to be interpreted as the order decided by, and given from, the higher branches of government with legally binding force. However, the legally binding force of such targets is applicable only within the administrative organizations and will not be extended to citizens, including the owners of enterprises.

Frequently, local governments would make regulatory rules or orders in order to implement targets on the basis of internal administrative targets decided by higher branch of government. Such regulatory rules or orders might include the administrative measures to impose negative duties or limit the rights and freedom of citizens, such as shutting down and/or removal of a factory, closing off the road from traffic, etc.

According to "Administrative Case Litigation Act of the People's Republic of China", the People's Courts shall not accept suits brought by citizens if the case involves administrative rules and regulations, or decisions and orders with general binding force formulated and announced by administrative organizations.

The regulatory documents such as regulation and orders with the aim to fulfill the allocated targets could be interpreted as the subject for the revocation of administrative dispositions in the Administrative Case Litigation Act. However, there has been no suit against the regulated documents nationwide, which have had specific or individual binding force to citizens.

As mentioned before, 31 provincial-level governments have submitted their "air pollution prevention and control responsibility documents" that contained AQIT with comprehensive work plan, but few work plans have implemented regulations enacted by local assemblies.

4.3.2.2 Regional Restrictions on Environmental Impact Assessment Approval

When a local government fails to achieve the allocated targets including AQIT, they will face punitive measure of sanctions from the EPA or the higher branches of local governments and get Regional Restrictions on Environmental Impact Assessment approval. If such sanction is imposed, any new projects on relevant region cannot proceed because of the restriction on the local government's authority to issue EIA approval.

For instance, Anyang, Henan Province, for which MEP imposed the restriction on issuing EIA approval for 223 days as part of efforts to fight against air pollution, had 118 projects suspended that had accounted for nearly 10 billion yuan (US$1.6 billion) in investment (Wang and Hao 2015). As one local environmental official said, "the restriction and the daily fines are equally important. Local governments get nervous when they are unable to get investment" (Wang and Hao 2015).

According to the study, 120 sanctions in total were imposed by July 2015, in which 48 sanctions were issued by MEP, 67 by provincial environmental authorities, and 5 by city-level authorities (Zhu et al. 2014). Since 2014, such restrictions have increasingly targeted the failures to meet total emission caps and environmental quality targets, while they used to target more in responses to breaches of EIA rules, failure to meet emission caps, incidents of lead poisoning, and heavy metal pollution (Zhu et al. 2014).

The study also suggested some risks in the abuse of power. First of all, even the revised environmental protection law stipulated that powers to impose restrictions would lie with the "environmental protection departments of the People's Government at province-level and above", though there have been the cases of lower branches of local governments imposing restrictions (Zhu et al. 2014).

Second, despite the law stipulating the sanctions to be applied only to "the construction projects, which create new emissions of key pollutants," focusing on the projects that would emit sulfur dioxide, nitrogen oxides, chemical oxygen demand, and ammonia, there have been cases where the restrictions would cover all construction projects (Wang and Hao 2015).

4.4 Conclusion

In this paper, we indicated how insufficient authorities given to China's Environmental Protection Bureaus and its lower status in governmental hierarchy caused leniency in the enforcement of environmental policies and hence the lower level of compliance in the past.

The Target Responsibility System and the management system for TRS process, such as Target Allocating Process and Target Achievement Process, have been important methods for China to overcome the problems in enforcing environmental policies.

Moreover, we analyzed the institutional background of introducing TRS and the challenges it faced. The legal analysis showed three main reasons for adopting the TRS method to ensure policy implementation: (1) strong command and supervisory authority of the central government over provincial governments, (2) strong binding force of FYP, and (3) greater power of the Communist Party of China (CPC) to control personnel affairs.

We pointed out that, especially because of heavy smog and increasingly severe air pollution developing in the "key regions," China introduced three regulatory action plans in 2012 to 2013, setting legally binding targets called Air Quality Improvement Targets, especially targeting PM2.5. Target Responsibility System would provide scoring system to evaluate target achievement, which would be reflected upon the evaluation of relevant governments and authorities and hence upon the personnel evaluation of individuals in charge. Such systems along with the introduction of four punitive measures would be likely to further incentivize the local governments, authorities, and personnel to achieve the allocated targets.

Moreover, China has implemented and improved the system of public engagement with sufficient environmental information disclosure while expediting the processing of the complaints and petitions reported by citizens.

About legal challenges to TRS, we clarified that even the legally binding force of targets would be applied only within the administrative organizations and would not be extended to the citizens, including the owners of enterprises. Moreover, there could be some risks of power abuse with the Regional Restrictions scheme of EIA approval. How to reform those abuses would be our next focusing points in the study of TRS in China.

References

The World Bank (2015) Gross domestic product 2014, PPP.

U.S Energy Information Administration (2015) China International energy data and analysis.

United Nations (2014) World Urbanization Prospects.

TsingHua University (2014) China and the New Climate Economy, The New Climate Economy (in Chinese).

Ministry of Environmental Protection of the People's Republic of China (MEP), General Administration of Quality Supervision, Inspection and Quarantine of the People's Republic of China (AQSIQ). National Ambient Air Quality Standard, GB, 3095–2012,

MEP (2012) Phase One Monitoring Implementation Plan for the New Ambient Quality Standard (in Chinese).

MEP (2015) 2014 Report on the State of Environment in China (in Chinese).

MEP (2014) 2013 Report on the State of Environment in China (in Chinese).

MEP, National Development and Reform Commission (NDRC), and Ministry of Finance (2012) 12th Five-Year Plan on Air Pollution Prevention and Control in Key Regions.

XinHua News (2013) 70 Chinese cities suffer serious air pollution in Q3 Xinhua Net edited. http://www.news/xinhuanet.com

The State Council of the People's Republic China (2013a) Atmospheric Pollution Prevention Action Plan - GuoFa [2013] No.37 (in Chinese)

MEP, NDRC, Ministry of Finance, Ministry of Industry and Information Technology of the People's Republic of China, Ministry of Housing and Urban-Rural Development of the People's Republic of China and National Energy Administration (2013) Implementation Rules of Action Plan against Air Pollution in Beijing, Tianjin, Hebei and Surrounding Areas - HuanFa [2013] No.104 (in Chinese)

China Economy Net News (2015) Chen Jining's new environmental policies (in Chinese), March 1th, 2015, Beijing (in Chinese) http://finance.ce.cn/rolling/201503/02/t20150302_4702242.shtml

Wang, Jin (Ed.), 2011. Rule of Law for 30 Years in Environmental Protection: Have We Made It? China's Environmental Protection Blue Book (1979–2010). Peking University Press, Beijing (in Chinese)

OECD (2006) Environmental Compliance and Enforcement in China. Paris

Ma, Xiaoying, Ortolano, Leonard (2008) Environmental Regulation in China: Institutions, Enforcement, and Compliance. Stanford University

Jin, Zhen, Kuramochi, Takeshi and Asuka, Jusen (2013) Energy and CO2 intensity reduction policies in China: Targets and Implementation. Global Environmental Research Vol.17.1 19–28, Tokyo

MEP (2013a) The MEP has signed target responsibility agreements for preventing and combating air pollution with 31 provinces. Beijing (in Chinese) http://www.gov.cn/gzdt/2014-01/07/content_2561650.htm

The State Council of the People's Republic China (2013b) Air Pollution Prevention and Control Action Plan. GuoFa [2013] No.37, Beijing (in Chinese)

The State Council of the People's Republic China (2014) Performance Assessment Measures for Air Pollution Prevention and Control Action Plan. GuoBanFa [2014] No. 21, Beijing (in Chinese)

Wang, Tao and Hao, Liting (2015) EIA regulation becomes key weapon in local fights to curb pollution. ChinaDiaLogue, Beijing https://www.chinadialogue.net/article/show/single/en/8105-EIA-regulation-becomes-key-weapon-in-local-fights-to-curb-pollution

Zhu, Xiao, Zhang, Lei, Ran, Ran, Mol, Arthur P.J. (2014) Regional restrictions on environmental impact assessment approval in China: the legitimacy of environmental authoritarianism. Journal of Cleaner Production 92 (2015) 100–108,

MEP (2014a) The Provisional rule for EMC to Implement the Interview. HuanFa [2014] No. 67, Beijing (in Chinese)

Time Weekly (2015), MEP is in its element: more than 20 cities were interviewed. TIME-WEEKLY.com, Beijing (in Chinese) http://www.time-weekly.com/html/20150804/30776_1.html

XinHua Net News (2015) Environment ministry targets fake air quality data. Apple 1th 2015, Beijing http://news.xinhuanet.com/english/2015-04/01/c_134117345.htm

China Council for International Cooperation on Environment and Development (CCICED) (2014) Progress in Environment and Development Policies in China (2013–2014) and CCICED Policy Recommendations Impact. Beijing

Institute of Public & Environmental Affairs (IPE) and Natural Resources Defense Council (NRDC) (2015) The 2014–2015 Pollution Information Transparency Index: sixth Annual Assessment of Environmental Transparency in 120 Chinese Cities. Beijing

IPE and NRDC (2015) The 2013–2014 Pollution Information Transparency Index: sixth Annual Assessment of Environmental Transparency in 120 Chinese Cities. Beijing

MEP (2013b) The Methods for Key State-Monitored Enterprise Pollution-Source Self-Monitoring and Information Disclosure (Trial). HuanFa [2013] No. 81, Beijing (in Chinese)

MEP (2014b) The notification for list of Key State-Monitored Enterprise in 2015. HuanFa [2014] No. 116, Beijing

EPA (2015) Report for resolution results of EPA "12369" complaint hotline on May 2015. Beijing (in Chinese) http://www.mep.gov.cn/gkml/hbb/qt/201508/t20150828_308870.htm

The Central Committee of CPC and the General Office of the State Council (2015) Accountability Methods on Environment Damage Caused by Leaders and Cadres of the Party and Government (Trial)". Beijing (in Chinese)

CIConsulting Net (2015) The governments mainly focusing in 2015: 31 provinces revised down GDP targets. Beijing (in Chinese) http://www.ocn.com.cn/chanjing/201502/difang251515.shtml

Li, Jinlei (2014) The Adjustment of Examination in Counties and Cities by Various Provinces, China is Ready to Say Goodbye to "Only GDP. China News, Beijing (in Chinese)

Chapter 5
An Analysis of China Coal Resource Tax Reform and Its Implications on Resource Saving

Yanmin He

Abstract As a traditional and feasible policy instrument, taxation is regarded for dealing with energy issues in politics and governance. This paper focused on the coal resource tax in China as an example of the energy tax and analysed the impacts of the tax system on fiscal revenue, resource conservation and energy savings for the period of 1994–2011. Based on this analysis, this paper discussed the goals and challenges of the 2011 coal resource tax reform, which changed the tax system from a volume-based system to a fixed-rate ad valorem one based on sales.

Keywords Resource tax reform • Coal resource • Environmental taxation • Energy structure adjustment • Fiscal reform

Abbreviations

GHG	Greenhouse gases
MOF	Ministry of Finance
SCE	Standard coal equivalent
CGE	Computable general equilibrium

5.1 Introduction

Along with China's rapid economic growth since it opened its economy to the outside world in 1978, the country's coal production and consumption have increased steadily. China became the world's leading coal producer in 1983 and the world's largest coal consumer in 1984. In 2014, China produced 3874 million tons of coal and consumed 3935 million tons in the same year, accounting for nearly

Y. He (✉)
Researcher of Institute of Economic Reseach, Kyoto University, Kyoto, Japan
e-mail: kaenmin@hotmail.com

H. Kitagawa (ed.), *Environmental Policy and Governance in China*,
DOI 10.1007/978-4-431-56490-4_5

half of global coal production and consumption.[1] This increase in coal production and consumption stems not only from economic growth but also from the economy's dependence on coal. China's Energy White Paper[2] (2007) states that the country's development will depend upon coal for many years, as China's energy mix are dominated by coal. Therefore, coal will remain a vital sector for China's energy security and growth stability over the long term. However, one major source of PM2.5 pollutants, formed from NOx, SOx, NH_3, VOCs, etc., is coal combustion. Extremely high levels of PM2.5 in key regions in China have triggered public concern and government action to restrict coal demand through the reform of resource taxation starting in 2011. However, China is already the top emitter of greenhouse gases (GHG), having surpassed the United States in 2006. As a result of increasing international pressure on China to control its emissions related to GHG, the Presidents of the United States and China announced their respective post-2020 actions and emission targets on climate change on 12 November 2014. China intends to achieve a peak in CO_2 emissions around 2030 and to put forth its best efforts to peak early. Under its new targets, China intends to increase the share of non-fossil fuels in primary energy consumption to around 20 % by 2030 through a set of mandatory targets and economic measures including taxation policies.

It is understood that taxation is a traditional and feasible policy instrument for dealing with energy issues. In particular, mineral resource taxation is used to adjust the cost gap and deal with externalities and market failure, leading to optimal coal consumption. This study focuses on the coal resource tax as an example of the mineral resource tax system. The coal resource tax was implemented in 1984 under the Draft Regulations on the Resource Tax, along with the natural gas resource tax and crude oil resource tax. The system levies and collects taxes on enterprises and individuals in China engaged in the extraction of coal.

Previous studies have mainly analysed the impact of the resource tax reform from a national perspective (Xi 2010; Cao et al. 2011; Fu 2012; Zhen 2012; Pei and Pu 2013). Although coal has been China's dominant source of energy, few empirical studies have investigated the effects of the coal resource tax on resource conservation and energy savings. Wang (2014) investigates the interest conflicts among coal mines, electricity companies and residents that resulted from the coal tax reform in China from a public policy perspective. He reveals that economic tools alone cannot result in good governance for coal, as the coal system includes such high stakes and involves so many players. Xie and Bai (2014) provide an overview of the transition of the coal resource tax system and discuss possible measures to readjust the taxation of coal. These measures include (1) shifting the tax burden of coal enterprises using a VAT rate adjustment; (2) linking the coal

[1]BP Statistical Review of World Energy 2015 (British Petroleum, June 2015) (http://www.bp.com/statisticalreview; accessed 11 August 2015).

[2]China's Energy Conditions and Policies (Information Office of the State Council, December 2007) (http://www.china.org.cn/english/environment/236955.htm; accessed 11 August 2015).

resource tax to a mining recovery ratio[3] and, eventually, to the recovery rate of the enterprise; and (3) having flexible rates for coal resources, which should be adjusted dynamically according to changes in the economic situation, among other factors. Shi et al. (2015) use a dynamic recursive computable general equilibrium (CGE) model by introducing a resource tax module to estimate the impacts of coal resource tax reform on China's economy, which indicates that the coal resource tax reform would improve China's energy structure adjustment and help reduce the total energy, especially coal in use and carbon emissions.

This study analyses the impacts of the coal resource tax system on fiscal revenue, resource conservation and energy savings for the period of 1994–2011, prior to the reform. Based on this analysis, this paper discusses the goals and challenges of the 2011 coal resource tax reform. In addition, we outline the recent status of coal resource tax reform in China after discussing the barrier to changing the tax system from a volume-based system to a fixed-rate ad valorem system based on sales.

The remainder of the paper is organized as follows. Section 5.2 introduces the characteristics of China's coal resource tax and its reform and provides a conceptual framework with which to study the impact of the tax reform. Section 5.3 investigates the effects of the coal resource tax before the tax reform, from 1994 to 2011. Section 5.4 discusses the effects of the 2011 coal resource tax reform as well as its goals and challenges. Section 5.5 outlines the current status of coal resource tax reform in China.

5.2 China's Coal Resource Tax: Transition and Characteristics

5.2.1 China's Coal Resource Tax Reform and Characteristics

5.2.1.1 A Short History of China's Coal Resource Tax System

Beginning in the mid-1950s, China introduced a centrally planned economy. Under this planned economy system, the government uniformly planned and directed prospecting for and the development of state-owned mineral resources. All developed resources were transferred to the government, from which the Ministry of Geology and Mineral Resources apportioned them, without compensation, to processing companies. For this reason, resource taxes were not levied on the extraction of mineral resources prior to 1984.

[3]Mining recovery ratio means the percentage of total tons of coal estimated to be recoverable from a given mine or area in relation to the total tonnage estimated to be in the demonstrated reserve base. It is a technology economic indicator to assess the superiority of mining technology and the work quality in an operation site.

In September 1984, a coal resource tax was introduced for coal mining companies by the Draft Regulations on the Resource Tax, along with taxes on crude oil, natural gas resources, metallic products and non-metallic products.[4] The coal resources tax was initially adopted as a progressive taxation. It was levied when a company's profit margin on the sale of taxable items exceeded 12%. However, tax exemption was granted on profit margin not exceeding 12%. For a profit margin between 12 % and 20 %, the tax rate increased by 0.5 % for each 1 % increase in profit margin. For a profit margin between 20 % and 25 %, the tax rate was increased by 0.6 % per 1 % of profit margin. However, for a profit margin greater than 25 %, the tax rate was increased by 0.7 % per additional 1 % of profit margin.

Following the enactment of the Mineral Resources Law (Presidential Decree No. 37), the Ministry of Finance (MOF) announced the 'Circular on taxing the resource tax on coal according to the volume-based method' (No. 291 of MOF, 1986) in September 1986. The coal resource tax thus changed from a progressive taxation system based on profit margin to one based on sales volume, which helped simplify the taxation procedures. After the change, only 30 state-owned mining companies were required to pay tax, with a specific tax rate for each coal mine.

In December 1993, the Provisional Regulations on Resource Tax (No. 139 of the State Council, 1993) and the Detailed Rules for the Implementation of Provisional Regulations on Resource Tax (No. 43 of the Ministry of Finance, 1993) were enacted. As a result, all coal-mining companies came to be levied, including non-state-owned mines. The tax rate was revised to between 0.3 and 5 yuan per ton. According to the Detailed List of Resource Tax Items and Value and the Table on the Grading of Mine Resources of Several Major Varieties attached to the Detailed Rules, mines were graded according to type and quality as well as production area. The amount of resource taxes to be paid was now based on the coal mine grade.

Since 1993, although the tax system has not been modified again, the central government has often adjusted the coal resource tax rate, depending on economic conditions and the development status of resources. A series of Circulars on Adjusting the Standard for Resource Taxes on Coal were implemented by the MOF and the State Administration of Taxation to raise the tax rate in the 30 autonomous regions and provinces across the country. The Circular on Adjusting the Standard for Resource Taxes on Coking Coal (Ministry of Finance and the State Administration of Taxation, No. 15 of the Ministry of Finance, 2007) was enacted in 2007 and applied a tax rate of 8 yuan per ton on coke coal, which was higher than the rate for general coal.

Through a resource tax reform pilot program centred in the Xinjiang-Uygur Autonomous Region beginning in 2010, the draft revision of the Provisional Regulations on the Resource Tax was enacted on 1 November 2011 for the first

[4]Shortly after the resource tax began to be levied with the promulgation of the Draft Regulations on the Resource Tax, MOF announced the 'Regulations on some Issues of Resource Tax' (No. 296 of the Ministry of Finance, 1984). The resource tax on metallic and non-metallic products was suspended, which meant that, until 1993, taxable resources were essentially crude oil, natural gas and coal.

time in nearly 30 years.[5] The resource tax rates before and after the reforms are shown in Table 5.1 Through the resource tax reforms, the taxation of crude oil and natural gas changed from a volume-based system to a fixed-rate ad valorem system. Although the coal resource tax remained a volume-based system, it was divided into a coke coal resource tax and other coal resource tax, along with an increase in the coke coal resource tax rate.

5.2.1.2 Characteristics of China's Coal Resource Tax

Before 1993, the central government planned and directed the bulk of the developed resources in China, resulting in a low regulated supply price for mineral resources such as raw coal. For this reason, many mining companies operated at a loss. Therefore, only the 30 major state-owned coal-mining companies, who had stable production, were classified as taxpayers. Furthermore, the coal mining processing plans were decided by the central government, not the mining companies. Consequently, the coal resource tax during this period was not an institutional design to supress the extraction of resources. Rather, it was used as a tax instrument to accommodate the income differences between state-owned corporations and to equalize the competitive conditions between enterprises derived from the differences in the resource endowment (Han and Yang 1985; Bao and Yang 1998; Guan and Su 1999; Guan 2001; Zhang 2000; Zhang and Zhou 2007).

Only after the launch of the Provisional Regulations on Resource Tax, by which all coal-mining companies became taxpayers, did China's coal resource tax become fully operational.

Table 5.1 Resource tax rates before and after 2011 resource tax reform

	Nov. 2011, before reform	Nov.2011, after reform
Crude oil	8–30 yuan/ton	5 % of sales volume
Natural gas	2–15 yuan/1000 m^3	5 % of sales volume
Coke coal	0.3–5 yuan/ton	8–20 yuan/ton
Other coals	0.3–5 yuan/ton	0.3–5 yuan/ton (unchanged)
Other non-metallic raw ores	2–30 yuan/ton	0.5–20 yuan/ton or 1000 m^3
Ferrous raw ore	0.4–30 yuan/ton	2–30 yuan/ton
Non-ferrous raw ore	10–60 yuan/ton	0.4–60 yuan/ton
Solid salt	2–10 yuan/ton	10–60 yuan/ton
Brine	8–30 yuan/ton	2–10 yuan/ton

Source: The Provisional Regulations on the Resource Tax; the Decision of the State Council on Amending the Provisional Regulation of the People's Republic of China on Resource Tax (No. 605 of the State Council, 2011)

[5]For detail of the pilot program, refer to He (2014).

The characteristics of the coal resource tax system beginning in 1993 were as follows:

- The tax was levied on the sale of coal at the upstream stage, mainly on the mining and processing of coal.
- The tax was implemented based on the volume of coal sales.
- Coal mines were divided into grades based on differences in the objective conditions, such as the quality of coal and the production region. Individual tax rates were prescribed per grade, within a certain range, and based on the statutory tax rate.

Therefore, starting in 1993, the coal resource tax levied was an extraction tax, as the companies were taxed based on the amount they extracted from the ground (also known as a severance tax), with the production weight as the standard used to calculate the tax payable. Generally, an extraction tax provides the government with a certain amount of revenue, while at the same time increasing the price of the resource and reducing the net earnings of the mining companies. As a result, the tax has the direct effect of restricting the volume of the resource being mined. In addition, the tax has an indirect energy-saving effect by improving energy efficiency.

5.2.2 A Conceptual Framework for the Coal Resource Tax Reform

This study analyses the effects and challenges of the coal resource tax for the period 1994–2011, prior to the tax reform, from three perspectives.

First is the effect of the increase in tax revenue from the coal resource tax. We examine this effect by analysing government income at the local government level.

Second is the effect on resource conservation. We discuss the resource extraction restraint effect of the tax on coal resource extraction.

Third is the energy-saving effect of taxation. We discuss how the tax decreased GDP-specific energy consumption, which is the volume of primary energy consumption as a percentage of GDP.

Finally, based on the results of our analysis, we investigate the goals and challenges of the 2011 coal resource tax reform.

At present, owing to insufficient relevant data on coal resource tax before 2001, this study uses data from the Chinese Statistical Yearbook from 2001 to 2013. We analyse data on 'coal production' from the Chinese Statistical Yearbook, as no official statistical data on coal resource extraction are available. We conduct our analysis using an ordinary least-squares regression in Stata 13.

5.3 Empirical Analysis of the Impact of the Coal Resource Tax Before the Tax Reform (2001–2011)

5.3.1 *The Funding Effect of the Coal Resource Tax*

Table 5.2 shows data from 2001 to 2011 on resource tax revenues, coal resource tax revenues, and local government revenues. Since 2001, coal resource tax revenues have accounted for about 15–29 % of the total resource tax revenue (per item, this places it second behind crude oil resource tax). The central and local governments jointly levy resource taxes. However, levying of the offshore oil resource tax, which is the portion that goes to the central government, has been suspended for a long time. Thus, in practice, all resource tax revenue goes to the local governments, effectively making it a local government tax. Although coal resource tax revenue constitutes about only 0.1–0.3 % of local governments' total tax revenue, showing that the amount is not so significant when compared to others taxes, but for local governments, it still serve as an important means to increase the annual revenue, especially in those regions with an abundance of coal resources.

5.3.2 *The Effects on Resource Conservation*

Theoretically, levying a coal resource tax causes resource prices to increase and mining companies' net earnings to decrease. As a result, it has the effect of restricting the volume of coal mined. In other words, coal resource tax and the

Table 5.2 Changes in coal resource tax revenue (2001–2011)

Year	Resource tax revenue (a) (10 thousand yuan)	Coal resource tax revenue (b) (10 thousand yuan)	(a)/(b) (%)	Local government revenue (d) (100 million yuan)	(b)/(d) (%)
2001	671,076	98,685	14.7	7803.30	0.13
2002	751,357	130,020	17.3	8515.00	0.15
2003	831,133	154,032	18.5	9849.98	0.16
2004	988,035	184,822	18.7	11,893.37	0.16
2005	1,422,030	381,211	26.8	15,100.76	0.25
2006	2,070,151	545,597	26.4	18,303.58	0.30
2007	2,610,244	729,983	28.0	23,572.62	0.31
2008	3,016,310	861,406	28.6	28,649.79	0.30
2009	3,382,165	960,570	28.4	32,602.59	0.29
2010	4,175,472	1,093,832	26.2	40,613.04	0.27
2011	5,958,700	1,260,421	21.2	52,547.11	0.24

Source: Compiled with data from Chinese Statistical Yearbook (2001–2012) (National Bureau of Statistics of China); Tax Year Book of China (2001–2012) (Editorial Board of China Tax Yearbook Bureau); Financial Yearbook of China (2001–2012) (China Financial Magazine)

volume of coal produced are negatively correlated. In addition to coal resource tax, gross domestic product (GDP) and the percentage of GDP provided by the secondary industry[6] (hereafter, 'secondary industry percentage') also affect the volume of coal produced. If the volume of coal produced is Y_t, the effective coal resource tax rate is TR_t, GDP is GDP_t and the secondary industry percentage is SI_t; then, the multiple regression function that explains the volume of coal produced is as follows:

$$\ln Y_t = \alpha_1 + \beta_1 TR_t + \beta_2 GDP_t + \beta_3 SI_t + \varepsilon_t.$$

Here, ln is the natural logarithm, ε_t is the error term and t is the year ($t = 2001$, 2002, . . ., 2011). When estimating using the volume of coal produced in China after 2001 (see Table 5.3), GDP, the secondary industry percentage, and the effective coal resource tax rate,[7] the conventional least-squares method results in the following regression formula.

Table 5.3 Changes in coal production and GDP, secondary industry percentage and effective coal resource tax rate (2001–2011)

	Coal production (SCE)	GDP (100 million yuan)	Secondary industry percentage (%)	Effective coal resource tax rate (yuan per ton) (SCE)
2001	105,029	109,655	45.15	0.94
2002	110,732	120,333	44.79	1.17
2003	130,992	135,823	45.97	1.18
2004	151,616	159,878	46.23	1.22
2005	167,786	184,937	47.37	2.27
2006	180,626	216,314	47.95	3.02
2007	192,136	265,810	47.34	3.80
2008	200,104	314,045	47.45	4.30
2009	212,280	340,903	46.24	4.53
2010	227,438	401,513	46.67	4.81
2011	247,394	473,104	46.59	5.09

Source: Compiled with data from Chinese Statistical Yearbook (2001–2012) (National Bureau of Statistics of China)

Note: China typically converts all its energy statistics into metric tons of standard coal equivalent (SCE)

[6]According to the Three Industrial Classification Regulation 2002 (National Bureau of Statistics of China) and the Three Industrial Classification Regulation 2013, China's secondary industry refers to the mining industry, manufacturing, power and heating, gas, water service industry and construction industry. It also includes the mining processing industry and the major industrial energy-intensive industries, which are the targets of the resource tax (http://www.stats.gov.cn/tjsj/tjbz/; accessed 11 Aug. 2015).

[7]Effective coal resource tax rate = total annual resource tax revenue (10,000 yuan) ÷ total annual production of coal (10,000 SCE ton).

$$\ln Y_t = -17.86 - 0.09TR_t + 0.78 \ln GDP_t + 2.95 \ln SI_t$$
$$(-8.15) \quad (-3.01) \quad (7.84) \qquad (5.74)$$
$$R^2 = 0.993, \overline{R^2} = 0.990, \; DW = 2.74, N = 11.$$

Here, R^2 is the coefficient of determination, $\overline{R^2}$ is the coefficient of determination adjusted for the degrees of freedom, DW is the Durbin–Watson ratio, and N is the number of samples. The numerical values in parentheses shown below the coefficients of the formula express the relevant t-values.

The results of the estimate indicated that effective coal resource tax rate, GDP, and the secondary industry percentage all have a significant effect on the volume of coal produced at the 5 % significance level. More specifically, the effective coal resource tax rate can have a negative effect on the volume of coal produced and positive effect on both GDP and the secondary industry percentage. In other words, an increase in the effective coal resource tax rate restricts the volume of coal produced.

5.3.3 *The Effects on Energy Savings*

After the Chinese market was opened to the world, the consumption of primary energy in China (measured as the volume of energy consumption as a percentage of GDP) declined significantly from 13.26 ton (SCE) at 1980 to 0.74 ton in 2011 (SCE).[8] The reason given for the decrease in the GDP-specific energy consumption was China's industrial structure (the GDP industrial composition) and, in particular, changes to the secondary industry percentage, which includes the mining industry and the major energy-intensive industries. In addition, as previously described, coal constitutes 70 % of all primary energy consumed in China. Since a coal resource tax levied on the mining and processing of coal causes resource prices to rise, it has the indirect effect of reducing GDP-specific energy consumption and promoting energy savings to a certain extent. Here, we investigate the policy effects of the coal resource tax in terms of the reduction in GDP-specific energy consumption, as well as its impact on the structure of the secondary industry.

The multiple regression function that explains GDP-specific energy consumption is as follows:

$$\ln EI_t = \alpha_2 + \beta_4 SI_t + \beta_5 TR_t + \upsilon_t,$$

where the GDP-specific energy consumption is EI_t, the secondary industry percentage is SI_t, the effective coal resource tax rate is SI_t and the error term is υ_t.

[8]Compiled from data from the 2012 Chinese Statistical Yearbook (National Bureau of Statistics of China).

When estimating using the secondary industry percentage (Table 5.3), the effective resource tax rate and China's GDP-specific energy consumption from 2001 onward (Table 5.4), we obtain the following regression formula:

$$\ln EI_t = -2.31 + 0.06SI_t - 0.15TR_t$$
$$(-2.90) \quad (3.51) \quad (-14.91)$$
$$R^2 = 0.969, \ \overline{R^2} = 0.962, \ DW = 1.703, \ N = 11.$$

The results of the estimates show that the secondary industry percentage and the effective coal resource tax rate have a significant effect on GDP-specific energy consumption at the 5 % significance level. While the secondary industry percentage has a positive effect on GDP-specific energy consumption, the effective coal resource tax rate has a negative effect. In other words, a higher coal resource tax rate means a lower GDP-specific energy consumption, which therefore promotes energy savings.

Based on the above estimates, we observe the following characteristics for the coal resource tax system between 1994 and 2011, prior to the reform:

(i) Levying a coal resource tax increased local government tax revenue.
(ii) The tax correlated negatively with the volume of coal produced. Therefore, an increase in the effective tax rate limited the volume of coal produced and conserved resources.
(iii) The results confirm that the tax had, to a certain extent, an effect on GDP-specific energy consumption. The effective coal resource tax rate correlated negatively with GDP-specific energy consumption. Thus, as a tax measure, it plays a role in improving energy savings.

Table 5.4 Changes in GDP-specific energy consumption and coal consumption (2001–2011)

	Energy consumption (10 thousand ton) (SCE)	GDP-specific energy consumption (ton/10 thousand yuan)	Coal consumption (10 thousand SCE)
2001	150,406	1.37	102,727
2002	159,431	1.32	108,413
2003	183,792	1.35	128,287
2004	213,456	1.34	148,352
2005	235,997	1.28	167,086
2006	258,676	1.20	183,919
2007	280,508	1.06	199,441
2008	291,448	0.93	204,888
2009	306,647	0.90	215,879
2010	324,939	0.81	220,959
2011	348,002	0.74	238,033

Source: Compiled with data from Chinese Statistical Yearbook (2001–2011) (National Bureau of Statistics of China)

Note: The volume of coal consumption here is calculated on the data of coal consumption in generating electric power

However, there are several other factors that need to be considered. Firstly, the volume of coal produced has increased continuously each year. Therefore, we cannot say that the coal resource tax has functioned sufficiently as an incentive to limit the mining and production of coal. Secondly, China's 11th five-year plan aimed to reduce GDP-specific energy consumption by 20 % between 2006 and 2010, and the 12th five-year plan (2011–2015) aims to reduce consumption by an additional 16 %. Therefore, we cannot ignore the energy-saving effects that have resulted from these policies, which are separate from the effects of the coal resource tax. Lastly, China is even now maintaining its industrial structure of consuming multiple types of primary energy, centred on coal, and China's GDP-specific energy consumption remains higher than those of developed countries. All of these additional factors should therefore not be overlooked when considering the effects of the coal resource tax.[9]

5.4 The Effect of the 2011 Coal Resource Tax Reform and Its Limits

In this section, we investigate how the systemic characteristics and problems of the coal resource tax system prior to the reform, discussed in Sect. 5.3, have changed as a result of the increase in the coal resource tax rate in the 2011 tax reform.

To begin with, Table 5.5 shows the coal resource tax revenue after the reform. Compared to 2011, revenue increased by around 1.5 billion yuan in 2013. Therefore, the reform continued to play a role in raising tax revenue. However, the table also shows that the percentage of total resource tax revenue provided by the coal resource tax declined, while the percentages provided by the crude oil and natural gas resource taxes increased. This was a consequence of the recent reforms, which changed the conventional volume-based system for crude oil and natural gas to a fixed-rate ad valorem system, greatly increasing their associated tax revenues.

In terms of an energy-saving effect, GDP-specific energy consumption in 2012 and 2013 were 0.70 ton per 10,000 yuan and 0.66 ton per 10,000 yuan, respectively, which was a further 5 % improvement over 2011.[10] On the other hand, the volume of coal being produced continued to increase from 2538.63 million tons (SCE) in 2012 to 2627.50 million tons (SCE) in 2013.[11] For example, if the coal resource tax reform had not been implemented and the volume of coal produced from 2012 onwards had increased at the same average rate as between 1992 and 2010, the estimated volumes for 2012 and 2013 would have been 2620.87 million tons (SCE) and 2689.41 million tons (SCE), respectively. These estimated production volumes exceed the actual production volumes. Therefore, the coal resource tax reform did,

[9] OECD (2013) Environment at a Glance 2013; IEA Energy Data Centre (2013); Energy Balances of Non-OECD Countries (2013 edition).

[10] Calculated from the 2012 Chinese Statistical Yearbook (National Bureau of Statistics of China).

[11] Ibid.

Table 5.5 Classified data of resource tax revenue (2010–2013) (100 million yuan)

	Total resource tax revenue	Coal	(%)	Crude oil	(%)	Natural gas	(%)	Others	(%)
2010	417.55	109.38	(26.20)	53.10	(12.72)	10.63	(2.56)	244.44	(58.54)
2011	595.87	126.04	(21.15)	137.37	(23.05)	28.74	(4.82)	303.72	(50.97)
2012	904.37	135.73	(15.01)	309.86	(34.26)	42.22	(4.67)	416.56	(46.06)
2013	1005.65	140.59	(13.98)	328.05	(32.62)	43.92	(4.37)	493.08	(49.03)

Source: Compiled with data from China Tax Year Book of China (2011-2014)(Editorial Board of China Tax Yearbook Bureau)
Note: "Others" refers to the tax revenue from Other coal, Other non-metallic raw ore, Ferrous raw ore, Non-ferrous raw ore, Solid salt and Brine

to a certain extent, restrict the volume of coal produced. However, given that the volume of coal produced continued to increase after the 2011 coal resource tax reform, the restricting effect was not that remarkable, for three reasons.

The first is that the effective tax rate is lower than the statutory tax rate. As shown in Table 5.1, the 2011 revision to the Provisional Regulations on Resource Tax raised the statutory resource tax rate for 'coke coal' from between 0.3 and 5 yuan per ton to between 8 and 20 yuan per ton. The statutory resource tax rate for 'other general coal' is between 0.3 and 5 yuan per ton, which has remained unchanged. In conjunction with this, as shown in Table 5.6, the enforced tax rate for coal resources in each province, autonomous region and directly controlled municipality has changed. Then, it was decided that, for each region, the enforced tax rate for 'coke coal' would be a uniform 8 yuan, to match the minimum statutory tax rate. Further, the enforced tax rate for 'other general coal' increased from between 0.5 and 1.6 yuan per ton (prior to the reform) to between 2 and 4 yuan per ton after the reform. However, this rate was still lower than the maximum statutory tax rate of 5 yuan per ton, and the average rate was even lower, at 2.8 yuan per ton. Finally, the effective tax rate per ton of raw coal, which prior to the 2011 reform was 3.6 yuan per ton, increased in 2012 to 3.8 yuan per ton but remained below the maximum statutory tax rate. Therefore, regardless of the fact that a higher statutory tax rate was introduced following the law revision, both the enforced tax rate and the effective tax rate in each region remained below the legal maximum rate. Therefore, the percentage that the coal resource tax constitutes of the market price of coal has declined, and the actual tax burden on coal mining and producing companies has become smaller. As a result, the coal resource tax has not been able to function sufficiently as an incentive to restrict the volume of coal produced.[12]

The second reason is that the coal resource tax is not linked to the price of coal. Here, we can verify the effect of the tax on the market price of coal using the

[12]According to the Coal Economic Operation Situation Summary of 2012 (China Industry News, 9 March 2013, A6), in many provinces, the coal resource tax amounts to about 2–8 yuan per ton, but the coal price reported by the China Bohai-Rim Coal Price ranges from 620–810 yuan per ton.

Table 5.6 Enforced tax rate for coal resources in provinces (yuan per ton)

	Before Nov.2011	After Nov.2011
Coke coal	N/A	8
Other coals		
Beijing	0.6	2.5
Hebei	0.9	3
Shanxi	1.6	3.2
Inner Mongolia	0.5	3.2
Liaoning	0.6	2.8
Jilin	0.6	2.5
Heilongjiang	0.8	2.3
Jiangsu	1.0	2.5
Anhui	1.0	2
Fujian	0.5	2.5
Jiangxi	0.6	2.5
Shandong	1.2	3.6
Henan	1.0	4
Hubei	0.5	3
Hunan	0.5	2.5
Guangdong	0.5	3.6
Guangxi	0.5	3
Chongqing	N/A	2.5
Sichuan	0.6	2.5
Guizhou	0.6	2.5
Yunnan	0.6	3
Shaanxi	0.5	3.2
Gansu	0.5	3
Qinghai	0.5	2.3
Ningxia	0.5	2.3
Xinjiang	0.5	3

Source: The Detailed Rules for the Implementation of the Provisional Regulations on Resource Tax; Wu and Lu (2014)

correlation between the Producer Price Indices for Coal Industrial,[13] $Price_t$ ($t = 2001, 2002, \ldots, 2012$), and the effective coal resource tax rate, TR_t, which results in the following formula:

$$\ln Price_t = 4.7067 + 0.0021TR_t$$
$$(92.36) \quad (0.01)$$
$$R^2 = 0.0353.N = 12.$$

The results of the estimate show that the effective coal resource tax rate does not have a statistically significant effect on the price of coal. The first reason for this is

[13] 2012 Chinese Statistical Yearbook (National Bureau of Statistics of China).

the mechanism used to determine the price of coal. China established a liberalized coal-trading market in 2009 after the central government planned fixed price period (1950 to 1992), partially liberalized period for coal other than coal for power generation (1993 to 2002) and the liberalized period with government participation for coal for power generation (2003 to 2008). This enabled trades to be made in the market itself. Theoretically, the current resource tax system should cause companies' production costs to increase, resource prices to rise, and the companies' net earnings to decline. This, in turn, should restrict the volume of coal produced. However, owing to the long period of government participation in coal for power generation, which constitutes more than 50 % of all coal consumed in China, companies' production costs are still not sufficiently linked to the market price of coal.

Furthermore, China's coal is unevenly distributed in the western interior region, including the Henan Province, Shanxi Province, Shaanxi Province and the Xinjiang-Uygur Autonomous Region. Also, major state-owned companies monopolistically use some of the mine railways.[14] In contrast, if small- and medium-sized coal mines, particularly those run by township and village enterprises, sell their coal on the coast, a long distance away, they pay high transportation costs. This causes their price competitiveness to decline. As a result, the coal supply becomes excessive in the areas surrounding the coal-producing regions, and the price of coal tends to remain lower than in coastal regions.

Furthermore, regional protectionism in coal-producing regions distorts prices. Some provinces, such as Shanxi Province, Henan Province, Shaanxi Province, and Anhui Province have implemented a policy called Mei Dian Hu Bao, which means a mutual protection between coal and power-generation enterprises. This policy prescribes that local power companies have preference when using coal produced within the region and only use coal from outside their region for needs not covered by locally produced coal. As a result, such regional protectionism measures have caused a variety of adverse effects on the market mechanism.

In addition to the abovementioned points, a major reason for the tax not being linked to the market price of coal is that the original coal resource tax was not a fixed-rate ad valorem system but instead was based on sales volume. Demand for coal has been growing due to the economic growth in recent years. As a result, the price of coke coal increased from 556.6 yuan per ton in January 2006 to 948.3 yuan per ton in October 2013, and the price of steam coal also increased from 204.9 yuan per ton in January 2006 to 339.1 yuan per ton in October 2013 (Fig. 5.1). Finally, the actual tax burden on coal mining for producing companies is not high enough to function as an incentive to restrict the volume of coal produced. Therefore, reforms were demanded that would change the system to a fixed-rate ad valorem system.

[14]For example, the Shenhua Group, the largest coal producer in China, built its own railway line connected to Qinhuangdao harbour, allowing a certain amount of monopolistic use. This system restricts use by third-party companies. We confirmed the actual situation in Shaanxi Province in a field survey from 3–4 March 2014.

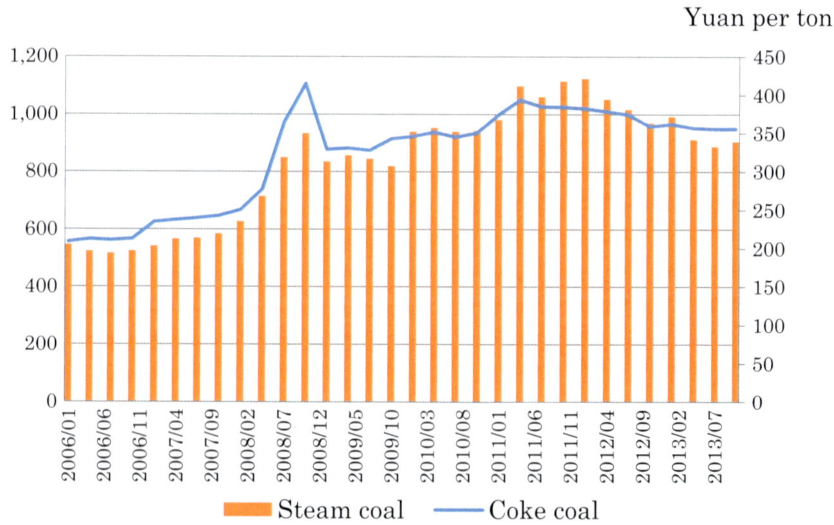

Source: Compiled with data from China Premium Database (CEIC Data)

Fig. 5.1 Changes in coke coal and steam coal price (2006–2013)

However, the resource tax system introduced in 2011 switched from a volume-based tax to an ad valorem tax on only crude oil and natural gas, leaving the volume-based taxation for coal, which constitutes 70 % of all primary energy consumed in China, the same as before. The reason given was that the different factors influencing coal production, such as the central government and local governments, coal companies, electric power companies, and consumers, all had different perspectives, and differences of opinion made it difficult to reform the tax system, as shown in Table 5.7 (Wang 2014).

Another issue is the existing fees involving coal. Due to historical reasons, Chinese coal producers pay 21 kinds of taxes including a corporation tax, VAT and a resource tax as well as fees and charges under various names, such as coal price adjustment funds, compensation fees for native minerals and fees for local economic development. These coal-related fees and charges, of which there are estimated to be more than 88 types (some with no legal or regulatory basis), account for around 40 % of a Chinese coal mine's total costs. Of that 40 %, one-third are fees and charges, most of which are double levied by the local administrative departments. One example is transportation related fees and charges, of which there are estimated to be 35 types.[15] In addition to a road funding fee, temporary land occupation fee, pollution treatment charge, road repair fee, surveying charge, specialized maintenance charge, service charge and others that are collected by the

[15]According to 'Ease burden on coal producers' (Economic Daily, 21 March 2013, A5) and 'Clearance the fees and adjust the taxes' (Chinese Coal Newspaper, 13 March 2013, M6).

Table 5.7 The actor and their position

Actor		Position
Central government	National Development and Reform Commission (NDRC)	A main macroeconomic management administration of China. It also takes responsibility for making coal policy and insists on the reform concerning about energy security, environment and potential over production risk in coal industry.
	State Administration of Taxation (SAT)	It charges the tax affairs in all countries including setting tax rate and making specific regulations. It insists on the reform concerning about fiscal revenue.
	State-owned Assets Supervision and Administration Commission of the State Council (SASAC)	It is the major stakeholder of all the state-owned coal companies. It insists on the reform concerning about energy security and potential over production risk in coal industry.
Local government	Coal production provinces	China's main coal production provinces concentrate in the central and western area, suffering a weak and undeveloped financial base. Coal production contributes these provinces' revenue under current coal resource tax regime. They are in favour of the main objectives central government wants to help them develop by coal tax reform.
	Coal consumption provinces	They are developed regions with larger population and more GDP without coal reserve. They oppose an increase in coal cost by tax reform.
Supply side of the market	Coal companies	They lie in the upstream of the production chain. China's coal industry has been fragmented by large state-owned companies, local state-owned companies and a large quantity of town and village coal companies. They firmly believe that government should alleviate the mass tax burdens and promote the market efficiency.
Demand side of the market	Electricity companies	They are the main consumer of coal and all are state-owned. As a main raw material of electricity generation, coal price increase will exacerbate their cost. Nonetheless, they cannot increase electricity price as a response because government controls the price. Electricity sector treats the reform as discrimination unless central government frees up electricity price.

(continued)

Table 5.7 (continued)

Actor		Position
	Residents	They are in the downstream of the production chain and use electricity for life. Many people support the reform in their value, because for a long time, the coal resource tax rate is too low to manifest the scarcity of coal. Consequently, coal mines can gain supernormal profit so easy that they neglect to protect the resources and environment. It is against the principle of social and ecological justice.

Source: Adapted from Wang (2014)

road sector, there are also exclusive railway maintenance charges, train car procurement fees, railway land rent, train car maintenance, railway rent, construction subsidies, platform rent and others collected by the railway sector and also a conveyance fee, warehouse fee, road usage fee, storage charge, and others collected by the waterborne transportation sector. Readjustment and elimination of coal-related fees and charges will be necessary to offset any increased taxes caused by the shift to an ad valorem tax system.

5.5 Current Status of Coal Resource Tax Reform in China

As described above, we found that the resource tax reform to the tax system has partially restricted the production volume of coal, but the effect is not particularly remarkable. The key reasons for this are that the coal resource tax is not linked to the price of coal, and the coal resource tax is based on sales volume. The new resource tax system introduced in 2011 switched from a volume-based tax to an ad valorem tax on only crude oil and natural gas, leaving the volume-based taxation for coal, due to the conflict of interest between stakeholders in the coal production industry and the backlash from the coal producers. After all, during the 2011 resource reform, amendments to the coal resources tax only raised the tax rate of coke coal, which was a compromise. But imposing a tax on production or consumption of coal can be used to adjust the cost gap and deal with externalities and market failure, leading to the optimum usage of coal. In this respect, an ad valorem coal tax will be very effective.

China is already the top emitter of GHG, having surpassed the United States in 2006. Although the restraint of coal use is thought to be the most effective way to reduce not only domestic greenhouse gases emissions but also global emissions significantly, it is feared that reducing coal consumption could have a negative impact on economic growth due to the economy's dependence on coal. As a result of increasing international pressure on China to control its emissions related to

GHG, China and the United States announced a bilateral deal, which was a commitment to work together along with other countries, to adopt a protocol at the United Nations Climate Conference in Paris in 2015. In connection with this deal, the State Council of China published the Energy Development Strategy Action Plan (2014–2020) that aims to reduce China's high energy consumption per unit GDP ratio through a set of mandatory targets and economic measures including taxation policies on 19 November 2014.

Thus, the MOF and SAT jointly issued the Circular on Implementing Coal Resource Tax Reform (No. 72 of the Ministry of Finance and the State Administration of Taxation) in October 2014. Under this circular, the resource tax on coal is now calculated on the basis of price rather than production volume across the whole country. Meanwhile, the regime of coal-related charges and special funds will be rectified. According to the joint circular, starting from 1 December 2014, the coal resource tax will be calculated on an ad valorem basis instead of based on production volume, with rates ranging from 2 to 10 %. The exact applicable rates will be proposed by provincial-level finance and tax departments to the provincial people's government for endorsement in consideration of factors such as the scale of the coal resources tax, the capacity of enterprises and the condition of coal deposits. Those regions with a higher tax burden will have their tax burden eased to some degree. The tax rates will be submitted to the MOF and the SAT for approval prior to official publication.

As the Circular referenced, accelerating coal resource tax reform will be able to promote the effective and intensive use of resources, protect the environment, and boost the construction of an eco-civilization. How the resource tax reform will impact companies' operations is dependent upon the tax rate and how vigorously fees are eliminated; both remain unknown. If tax increases are offset by eliminating coal-related fees, then the reforms could have a minimal impact on coal producers and energy consumers. However, if tax increases are not offset by eliminating fees, then reforms might result in less efficient coal producers going out of business. It is necessary to continue to monitor implementation and provide an analysis on how resource tax reform affects the coal industry and the Chinese economy.

On the other hand, a cap on annual coal consumption was introduced by the Energy Development Strategy Action Plan, so it should be kept below 4.2 billion tons until 2020, with the main coal consumption reduction to be achieved in regions around Beijing, the Yangtze River Delta and the Pearl River Delta, the three biggest urban clusters in China. In some Chinese provinces, coal consumption caps were mandated, with the aim of helping China peak its coal consumption by 2020.

Therefore, a mix of economic incentives and regulatory measures is being considered as a means for controlling coal use and accelerating the replacement of coal with energy efficiency and cleaner energy sources, which will fundamentally help China achieve its long-term economic, environmental and climate goals, and the policy mix has been having actual effects. The Chinese government recently announced that 2014 saw a 2.9 % reduction in coal consumption in absolute terms.[16] However, whether the new coal resource tax can help explain the reductions and whether the reductions can be sustained must be studied further.

5.6 Conclusion

In this paper, we verified China's experience on coal resource tax reform and revealed the effects of coal resource tax on the increase in tax revenue, on the conservation of coal resources and on energy savings. We clarified that levying a coal resource tax increases local government revenue by increasing tax revenue, and the coal resource tax restricts the volume of coal produced, thus conserving coal resources and improves energy efficiency.

References

Cao, Aihong and Han, Botang, Qi, Antian (2011), 'The Policy Research on the Resource Tax Reform in China(中国资源税改革的政策研究)', China Population, Resources and Environment, 21, pp.158–63.

China Financial Magazine (eds.) (2001-2012), Financial Yearbook of China (中国财政年鉴), CN: Financial Magazine of China.

China Premium Database (CEIC Data)

Editorial Board of China Tax Yearbook Bureau (eds.) (2001-2012), Tax Year Book of China(中国税务年鉴), CN: China's Tax Press.

Fu, Li. (2012), 'China's Resource Tax Reform (我国资源税改革问题研究)', Forestry Economics, 5, pp.117–19.

Guan, Fengjun (2001), 'Theoretical Analysis of the Resources Tax and Compensation(资源税和补偿费理论辨析)', China Geology & Mining Economics, 8, pp.1-3.

Guan, Fengjun and Su, Xun (1999), 'Several Opinions about the Mineral Resources Compensation (关于矿产资源补偿费的几个观点)', Resource and Industry, 8, pp.12–13.

Han, Shaochu and Yang, Yimin (1985), 'Several Opinions about the Resource Tax Levy Problem (对开征资源税问题的一些认识)', Public Finance Research, 2, pp.56–61.

He, Yanmin (2014), 'Effects and issues of the 2010 resource tax reform in Xinjiang', Kreiser, Larry, Lee, Soocheol, Ueta, Kazuhiro, Milne, Janet E., Ashiabor, Hope (eds.), Environmental taxation and green fiscal reform: theory and impact, UK: Edward Elgar, pp.215–226.

National Bureau of Statistics of China (eds) (2001-2012), China Statistical Yearbook(中国统计年鉴), CN: China Statistics Press.

Pei, Xiao and Pu, Zhizhong (2013), 'An Empirical Investigation of China's Resource Tax(我国资源税的经济效应分析)', Finance and Accounting Monthly, 8, pp.29–32.

Shi, Jiarui et al. (2015),'Impacts of coal resource tax reform on China's economy based on a CGE model', Systems Engineering — Theory & Practice, 35(7), pp.1698–1707.

Wang, Dong (2014), 'Analysis on Conflicts of China's Coal Tax Reform', International Journal of Energy Economics and Policy, 4(1), pp.108-116.

[16]CAAC Clean Air Management Report 2015 (Clean Air Alliance of China, 31 July 2015) (http://en.cleanairchina.org/product/7228.html, accessed 11 Aug. 2015).

Wu, Xinwen and Lu, Wutao (2014), 'A Review and some suggestions of coal resource tax reform transition in China(我国煤炭资源税改革历程回顾及建议)', Economic Tribune, 522(1), pp.108–112.
Xi, Xiaojin (2010), 'An Analysis on the Economic Effects of Resources Tax in China(我国资源税经济效应实证分析)', Co-operative Economy & Science, 399, pp. 92–4.
Xie, Xin and Bai, Lu (2014),'Discussion on the Achievements and Problems of Coal Resource Tax Reform and Counter-Measure Research(我国煤炭资源税改革效果、问题及对策研究)', Natural Resource Economics of China, 316(3), pp.22–26.
Bao, Ronghua and Yang, Hulin (1998), 'The existing problems and improvement measures of China's mineral resources tax and charges(我国矿产资源税费征收存在的问题及改进措施)', Geological Techno economic Management, 20(4), pp.20–22.
Zhang, Wenju (2000), 'Evolution and Development Orientation of Property Rights System of mineral Resources in China(我国矿产资源财产权利制度的演化和发展方向)', China Geology & Mining Economics, 1. pp.1–10.
Zhang, Jugang and Zhou, Jiguang (2007), 'Theoretical and Practical Study on Mineral Resource Tax in China(我国矿产资源税问题的理论与实践研究)', Journal of Shijiazhuang University of Economics, 30(4), pp.57–60.
Zheng, Wen (2012), 'An Empirical Analysis on the Impact Factors of Resource Tax(我国资源税影响因素的实证分析)', Financial Supervision, 11, pp.69–70.

Chapter 6
Environmental Litigation and External Influence from Outside the Court in the PRC: A Case Study of Zhang Changjian et al. v. Rongping Chemical Plant

Jiro Sakurai

Abstract In this chapter, I will examine how some political bodies impact the judicial courts where environmental litigations take place. The focus is on the political bodies that have the most substantial influence on a local court if certain political bodies, such as the upper class court, the local Committee of the CPC, or the Environmental Protection Agency of the upper class local government, offer suggestions on the same case.

By looking through the whole dispute process, this paper concludes that the decisions regarding an environmental dispute made by the local Committee of the CPC could have a substantial impact on the decisions of the local court, while the EPA of the upper level local government or the upper level local court could be disregarded by the local court.

Keywords Environmental litigation • Environmental dispute • Lawsuit • Judgment • Court • Case study • Judicial • Indemnification

Abbreviations

CPC	Communist Party of China
EPA	Environmental Protection Agency
CCTV	China Central Television

J. Sakurai (✉)
Department of Chinese Studies, Kobe City University of Foreign Studies, Kobe, Hyogo, Japan
e-mail: sakurai133@gmail.com

H. Kitagawa (ed.), *Environmental Policy and Governance in China*,
DOI 10.1007/978-4-431-56490-4_6

6.1 Introduction

With its rapid industrial development, public damage by environmental pollution can be found almost everywhere in China (Economy 2004). However, most of the victims are ignored without any compensation, or they are forced to settle for little compensation (van Rooij et al. 2012).

In this chapter, an environmental lawsuit at Pingnan County-level city (hereafter "Pingnan County") in Fujian Province will be examined. Despite the difficult conditions in China, more than 1700 pollution victims in Pingnan County filed a lawsuit and gained indemnification. Therefore, this case has drawn attention not only in China but also from abroad as one of the biggest environmental class actions in China. This case was also chosen as one of the "ten most significant impactful litigations of 2005 in China," and it was also chosen as one of the eight representative and influential cases of the period by China Dialogue on July 26, 2011.

While 10 years have already passed since the final judgment of this suit, and there has already been commentary on this case in several articles (Wang 2007, Stern 2011, 2013), there are still some remaining points to be elaborated on in this case.

First, the influence of some political bodies on the environmental litigation will be examined. As regards this Pingnan case, some political bodies have held meetings to settle this conflict. The first meeting that was held to settle the conflict was held by the local government of Pingnan County, and then the local Committee of the Communist Party of China (hereafter "CPC") held a special meeting about this issue and confirmed some decisions. Then, the Environmental Protection Agency (hereafter "EPA") at the provincial level held a meeting and compiled some suggestions for the process of the environmental impact assessment. But the suggestions compiled by the provincial EPA have seemingly not had any impact on the local court, while the unanimous decisions made by County Committee of the CPC have had a substantial impact both on the local court and on the local government in the process of the litigation.

Another point to be elaborated on here is the question of which political body will have more substantial influence on a local court if certain political bodies, such as the upper class court, the local Committee of the CPC, or the Environmental Protection Agency of the upper class local government, give the court some suggestions about the same case. For instance, there was an illegal intervention by the district court during the execution process of the judgment under the leadership of the local Committee of the CPC, but the upper class court has not stopped that intervention even though they were notified about it.

This article discusses these two points by elaborating on each person's or group's behavior, decisions, or suggestions regarding this pollution-oriented dispute throughout the whole process, which includes the judgments of the local courts and the plaintiff's movements regarding this issue.

6.2 Summary of the Case

Pingnan County is located in the northern part of the Mingdong Mountains in Fujian and was designated as a poor county. It was Fujian Rongping Chemical Co., Ltd. (福建省(屏南)榕屏化工有限公司) (hereafter "Rongping Chemical") that caused the pollution. Rongping Chemical had received the support of some of the leaders of the local governments and the local CPC since the company was invited by Pingnan County in 1992. The tax revenue from Rongping Chemical accounted for one-third of the annual income of Pingnan County.

Hexavalent chromium and chlorine were present in the drainage and the solid wastes from Rongping Chemical, and the effluent gas contained chlorine, too. These substances mainly damaged the health of the villagers around Rongping Chemical and affected economic activities, such as farm products, fruit trees, or bamboo trees for commerce. In addition, it was confirmed that waste dumped in a western valley by one factory had not yet undergone any processing when it was checked in the field in March 2010.

The victims around the factory appealed to the local government and the local CPC for intervention in this pollution problem many times, especially beginning in 1999 when Rongping Chemical expanded their plant, leading to more widespread damages, but the government only paid small sums as compensation for the damaged farm products, and more critical to the neighboring farmers, the pollution did not stop. Therefore, the victims decided to file a lawsuit in 2002. This case drew the attention of many media outlets, including China Central Television (hereafter "CCTV"), and both the first trial and second trial admitted a portion of the claim of indemnity (approximately 6 % of the plaintiff's request) and also admitted the claim of pollution suspension.

However, the judgment concerning the pollution suspension has not been carried out, and the natural environment around the factory has not yet been restored, either. When the author visited Pingnan County in March 2010, there was ongoing construction to add new facilities beside Rongping Chemical's existing plant.

6.3 The Meeting of the Government of Pingnan County on July 20, 2000

6.3.1 *Summary of the Meeting*

This meeting was organized to solve the pollution issue occurring in Pingnan County, and the minutes were recorded. Some points mentioned as follows are based on the minutes of the meeting, "Minutes of meeting No. 54 (专题会议纪要五四)."

The county government vice-president convened the meeting, and 18 people participated, including a county government advisor, county government executive members, the county government staff, the Pingcheng Township president, and a representative of the villagers of Xiping village. The minutes confirm that the participants agreed with the following two points.

The first point was about the location of Rongping Chemical, and they admitted that was a rational decision. The second point was about compensation, and they also admitted that compensation based on a certain standard was necessary because Rongping Chemical had adversely affected neighboring farm products. Based on these agreements, they reached the following specific decisions in the meeting.

About the adverse effects that the neighboring environment suffered as a result of the pollution by Rongping Chemical, the government would expend RMB 40,000 (if some portion is left over, then it should be returned to the budget, and if it is not sufficient, the amount should be supplemented). In addition, Xiping village would invite in an environmental monitoring company for an inspection, during which the company would evaluate the status of the pollution, calculate the amount of losses, and submit a pollution decontamination plan.

About the compensation for the farm products that already suffered damage, the government would pay an amount of compensation based on a compensation cost calculation standard for the farm products in the range that the government authorized.

The facility for disposal of the waste should be offered for immediate use.

The county government economic committee and the Xiping village party committee of the CPC should sign an agreement.

6.3.2 *Argument of Rongping Chemical Against the Decision of the County Government*

On August 2, 2 weeks after the county government's decisions in this meeting, Rongping Chemical sent the document "Ping Lian Hua 2000, No.41" (屏联化(2000)41号) of not agreeing to the county government's decisions to the county government. The claims of Rongping Chemical listed in "Ping Lian Hua (2000) 41" are as follows.

Rongping Chemical does not pollute the local natural environment. The claims of the villagers are not based on any scientific evidence. Rongping Chemical has protected the local natural environment positively and consistently.

Rongping Chemical has invested a large amount of money in environmental protection facilities and has been strictly observing emission standards. Unusual weather might be the main factor in the poor harvests of local farm products.

Rongping Chemical invested one million yuan or more to build waste disposal facilities; however, it cannot yet use them based on the objections of the local people. That depends on other circumstances, such as that Rongping Chemical has stored industrial discharge on a road to the nearby sewage treatment facilities.

Waste piled up in this way may pollute river water, and our company is very concerned about it. Even if pollution were to occur because of this industrial waste, it would be difficult for our company to take responsibility.

Based on these records of the Pingnan County government and Rongping Chemical, we know that when the conflict brought about by the pollution had just occurred, the local government admitted that Rongping Chemical's influence on pollution had damaged farm products, and they also recognized the need for compensation in handling this conflict regarding the pollution issue. On the other hand, Rongping Chemical flatly objected to the decision of the Pingnan County government about the pollution damage mentioned above and refused to take responsibility for the pollution damage.

6.4 Meeting of the Pingnan County Committee of CPC on March 20, 2002

6.4.1 Summary of the Meeting

The secretary of the Pingnan County Committee of the CPC convened this meeting, and the participants were limited to the members of the Pingnan County Committee of CPC. When this meeting was held, letters and visits of complaints by inhabitants had reached the central government and the Central Committee of CPC and not just the factory and the local government. As a result, an e-mail from Zhu Rongji, the prime minister at that time, reached one of the victims on January 12, 2002. This meeting of the County Committee of the CPC was held with this intensifying dispute going on in the background.

The following points about this meeting are based on "The Pingnan County Party Committee (Minutes of Meeting 3)" (中共屏南县委《专题会议纪要》(三)).

The participants were party committee members of Pingnan County, including the Pingcheng Township president, county government officials, and the party committee secretary chief of Rongping Chemical.

At first, in the meeting, the participants agreed to the following five points.

"The final goals"

The final goals of this meeting include the development of the company, the actual profit of villagers, environmental improvement, and a good relationship between the factory and Pingcheng Township.

"Propriety of the factory invitation decision"

Rongping Chemical was invited by leaders of the party government of Pingnan County at the time. In addition, leaders of the party government of Fujian Province,

Fuzhou City, and Ningde City supported this invitation decision. This meeting therefore denied the rumor that "the factory is garbage moved from Fuzhou City to the village of Pingnan," which had been spread by some villagers.

"Local contributions of Rongping Chemical"

Rongping Chemical increases the wealth, income, and employment rate of all citizens of the prefecture, contributes to the industrial development, and contributes to the economic and societal development of Pingnan County.

"The environmental pollution to neighbors of the factory"

The factory of Rongping Chemical definitely polluted some neighboring environments and definitely also resulted in damage to the health of the people of the Xiping village.

"Attention on people causing disturbances"

The Pingnan Party Committee welcomes opinions or suggestions that are submitted through normal and lawful channels. On the other hand, when certain heartless people destroy social stability by activities that are illegal, we consider such acts to be criminal activities that must be dealt with strictly and firmly.

The facts from this meeting can be confirmed as follows.

First, the Pingnan County Committee of CPC recognized the outbreak of the damage caused by the pollution from Rongping Chemical at this stage. Nevertheless, they justified the invitation to Rongping Chemical and identified the development of Rongping Chemical as their final goal. Then, they considered acts against the pollution of Rongping Chemical to be criminal, for such acts would "destroy social stability."

6.4.2 *Decisions in the Meeting*

Based on the abovementioned common agreements, the following decisions were achieved at the meeting.

"The reinforcement of measures against pollution by the factory"

Rongping Chemical should adopt effective general measures for pollution, and the Environmental Protection Agency should provide overall counsel for the pollution measures of Rongping Chemical. During monitoring, we request the participation of representatives of the People's Congress, representatives of the Political Consultative Conference, leaders of Xiping village, and representatives of the villagers.

"The greening around the factory"

The Pingnan County agriculture office, the Pingnan County forestry office, and the Pingcheng Township government should educate the people of Xiping village about planting farm products resistant to chlorine. And they should also help with agricultural production and environmental tree planting.

"The investigation on the cause of the illness"

About the health problems of the people, the Bureau of City Appearance, Environment, and Sanitation (市容环境卫生局) are carrying out the investigation of the recent cancer patients in Xiping village. On the other hand, Rongping Chemical should carry out a full-scale investigation of the staff of its own factory and the staff of the Fuzhou First Chemical Company (福州一化化学品股份有限公司) and should make the incidence of cancer in the staff clear.

"Compensation for damages"

Rongping Chemical should compensate the victims for damages if it becomes clear that the cause of the damage is the operation of Rongping Chemical. However, if someone is seducing people through falsehoods or agitating the people to cause a disturbance, the agencies concerned should punish them severely.

"Publicity and explanation"

The county government should produce certain documents with details about the effects and socioeconomic aspects of Rongping Chemical's invitation to correct the misunderstandings of the people. In addition, the county government should publicize the significance and positive effects of the factory invitation to the people and then explain it to the upper party organization properly. The leaders of Pingnan County should hold a roundtable conference to hear the opinions of the people. Also, the leaders of Pingnan County will strengthen control on the media and supervise it so that it will be really objective and fair and report the correct news.

"To maintain social stability"

The people of Xiping village must act according to the law and plead through legal channels. If a few people commit unlawful acts, the political law section will investigate those illegal acts and deal with them determinedly according to the law.

"The mutual prosperity of the factory and the village"

Rongping Chemical gives priority to Xiping village in contracts regarding the work of construction, just-in-time employment, and material transport. In this way, the factory and the village should try to build a relationship to develop together and prosper together.

At the time when this meeting was held, many victims had started a campaign to collect signatures and donations to appeal to a trial. However, the local county government suspended the signature campaign forcibly, and the donations gathered were confiscated. Moreover, in the process of this clampdown, the local police not only destroyed the signboard for signature campaigns and threatened the participants but also resorted to violence against some participants, too. We can tell that the signature campaign and the donation activities to appeal to a trial were considered acts that could "destroy social stability" and were not considered to be the "legal routes" that were mentioned by the decisions of the County Committee of the CPC. The fact that even a signature campaign and the donation activity were targeted for the clampdown shows us the difficulties the environmental victims faced in the steps toward lawsuits in China.

6.5 Public Hearing Hosted by the Fujian Provincial Environmental Protection Agency on November 20, 2002

6.5.1 *Summary of the Meeting*

This public hearing was hosted by the Fujian Provincial Environmental Protection Agency (hereafter "Provincial EPA"), and it was held as a part of the environmental assessment process regarding the authorization for Rongping Chemical to proceed with a production capacity reinforcement project. After this hearing of public opinions, the Provincial EPA approved the decision to allow Rongping Chemical to start a capacity reinforcement project.

In the background during when this hearing was held, the fact that the State Environmental Protection Administration (SEPA, now the Ministry of Environmental Protection) announced 55 unlawfully polluting companies in July 2002 and Rongping Chemical was included among them drew much attention. The hearing of public opinions in the environmental assessment process is provided as one of the means to solicit public opinions by the environmental impact assessment law enforced in 2003, which is not set as compulsory. So there must be some reason why a hearing of public opinions without legal obligation was held expressly.

The Provincial EPA compiled various suggestions and opinions given by the participants of the hearing as "acknowledgment of public hearing minutes" (听征会纪要). And this "acknowledgment of public hearing minutes" was sent to the key parties. The following descriptions about the hearing that the Provincial EPA hosted refer to this "acknowledgment of public hearing minutes." Suggestions and opinions summarized in the minutes are discussed in 5.2.

6.5.2 *Suggestions and Opinions Summarized by the Provincial EPA*

The following points are the suggestions and opinions summarized by the Provincial EPA.

"Illegally disposed of waste"

Rongping Chemical should appropriately dispose of the chromium waste they dumped illegally by the end of this year.

"Environmental protection director system"

Rongping Chemical should establish an environmental protection director system, and Rongping Chemical also should exchange opinions with representatives of Xiping village and Houlong village regularly in the factory.

"Environmental monitoring"

The county government should establish an environmental monitoring organization to monitor the discharge conditions of Rongping Chemical and its influence on the neighboring environment. The county government should allocate contracted staff in the environmental monitoring organization and should assign a budget. In addition, the county government should disclose information about the conditions of Rongping Chemical's discharges to the people. The right to know and the supervision of the people will in this way be guaranteed.

"Hygiene protection isolation zone"

The county government should urge the sections concerned to establish a hygiene protection isolation zone around the Xiping industrial district. Moreover, the county government should find an appropriate waste disposal site as soon as possible.

"Continuous monitoring system"

Rongping Chemical should set up a continuous monitoring system for the hexavalent chromium in the wastewater and install warning facilities for chlorine gas discharged into the atmosphere.

"Effluent gas purification facilities"

Rongping Chemical should install effluent gas purification facilities on the electrolysis product line and should ensure the chlorine in the effluent gas reliably meets the standard.

"Reinforcement of the management of the factory itself"

Rongping Chemical should strengthen the operation management of its product lines and pollution processing facilities to observe the emission standards of the country. Rongping Chemical should strengthen the education for its staff and should also accept supervision by the citizens in a positive manner.

"Reinforcement of the management of the factory"

Pingnan County should strengthen the supervision management of Rongping Chemical.

"Tree planting"

Rongping Chemical should plant some fruit trees and some bamboo trees around the factory.

In "the public hearing minutes" mentioned above, the final agreement was not reached unanimously, unlike the decisions of the county government meeting or the County Committee of the CPC. So "the public hearing minutes" are just documents prepared by the Provincial EPA, which aimed to promote the appropriate steps for the persons concerned. Hence, the opinions of local villagers are reflected more in these "public hearing minutes" than in the abovementioned meeting decisions. Therefore, through the examination of these "public hearing minutes," where the intentions of the villagers at this time can be understood more exactly, monitoring Rongping Chemical in order to stop the pollution damage seems to be the goal of the villagers. It is also significant to find no mention about the amount of the compensation in these minutes.

When this public hearing was held, the pollution victims had already filed lawsuits to demand pollution suspension and compensation for damages against the factory, which took place on November 7, 2002.

6.5.3 Investigation by the Provincial EPA

On April 12, 2003, China Central Television (CCTV) broadcasted a 20-min documentary in the program called "News Investigation" (新闻调查), which focused on pollution issues associated with Rongping Chemical. After this broadcast, the vice-president of Fujian Province ordered the Provincial EPA to conduct an investigation about this pollution problem, and then a special team led by the Provincial EPA's vice head, who was appointed as the team leader, was organized. This special team carried out an unannounced investigation at Rongping Chemical, and representatives of the county government and Rongping Chemical and representatives of the villagers were invited to the discussion held by the County EPA.

Then, on August 13 of the same year, SEPA published a list of ten unlawful environmental cases, and Rongping Chemical was included in this list. Indeed, according to a reporter for the "Prosecution Daily" (检察日报), who interviewed locals during this period, he smelled an acrid odor around the factory and saw a pale, yellow gas like a fog that wafted up from the factory.

6.6 Examination of the Judgment

6.6.1 Basic Information About the Suit

In March 2002, the pollution victims applied to the Center for Legal Assistance to Pollution Victims (CLAPV: 污染受害者法律帮助中心) for legal support through the introduction of a journalist. CLAPV is a research and service center that a professor named Wang Canfa of the China University of Political Science and Law established at his own University in Beijing. It has provided legal assistance to many pollution victims from all around China since 1999.

Professor Wang Canfa decided to support the victims of this case and himself became the defense lawyer along with an exclusive lawyer from CLAPV. He filed the petition for the suit based on the documents handed over by the victims and submitted it to the Ningde district court on November 7 of that year.

The plaintiffs of this suit were 1721 inhabitants of Pingcheng Township, most of whom were farmers, and the defendant was Rongping Chemical.

The requests in the suit were the following:

Stop the illegal discharge of pollution

Compensate the plaintiffs (approximately 10,290,000 yuan in total) for economic damage such as agricultural and forestry products

Compensate the plaintiffs (approximately 3,020,000 yuan in total) for mental damage

Dispose of the waste piled up in the factory and on the mountain behind the factory appropriately

The first trial was carried out in the Ningde prefecture Intermediate People's Court (中级人民法院), and the second trial was examined in Fujian Province's Higher People's Court (高级人民法院). In general, in the Chinese Civil Procedure Law, the court should announce a judgment within 6 months from a suit, but the first trial judgment in this case was April 2005, so it took approximately 2.5 years. So the duration of this trial was about five times the duration prescribed by the act, but oral proceedings were only opened twice during this period.

In addition, the second trial judgment was November 2005, which was 7 months after the first trial judgment, and oral proceedings were only opened for this trial once.

6.6.2 *Judgments of the Courts*

"First trial judgment"

The points of the first trial judgment were as follows.

(i) The court ordered the defendant to stop the illegal discharge of pollution.
(ii) The court also ordered the defendant to pay the plaintiff 249,763 yuan as compensation for the damage to the agricultural and forestry products.
(iii) The court dismissed the claim for payment for mental damage.
(iv) The court ordered the defendant to dispose of the wastes appropriately within 6 months from the date when a disposal method was established.

The court judgment ordered "the defendant to stop the illegal discharge of pollution," but the specific content of the "illegal discharge of pollution" was not identified in the decision. Only after looking into the reason for the judgment do we find the scope of the "illegal" activity here to be limited to the pollution damage to the agricultural and forestry crops. For in the reasons of judgment, the relevancy between the health damages and the pollution by the factory was not accepted; however, the relevancy between the damage to the agriculture products and the pollution by the factory was accepted.

On the other hand, the reason of the judgment does not deny the claim of the defendant who observed the standard of the country. Therefore, it can be understood that "to stop the illegal" activity means to prevent the damage to farm products and not just observe the effluent standard. But it is still not clear how and to what extent the defendant should prevent pollution damage to the agricultural and forestry crops.

As regards the amount of the compensation for the agricultural and forestry products, it was approximately 2.4 % of the plaintiffs' bills, and this was the sum that the defendant insisted on in settlement talks carried out in the court, according to the notes of the plaintiffs taken in court.

This judgment was not sentenced at the court but was delivered to the person concerned, and here we find a strange gap between the date written on the judgment and the date of the delivery. The date on the judgment is stated as "April 15, 2005," so this judgment can be understood to become effective on that day; however, it was May 15 when the judgment papers were delivered to the representative of the plaintiff, which can be confirmed by the date on the envelope from the court. Therefore, it took approximately 1 month for the judgment to be delivered to the plaintiff with no explanation nor reason why such a period was needed for the delivery of the judgment to the plaintiff.

Similarly, November 16, 2005, is stated as the date after the signature of the judge in a second trial decision, but it is December 22 when the judgment was delivered to the representative of the plaintiff. Just as in the first trial judgment, approximately 1 month was needed for the second trial decision to be delivered.

"Second trial judgment"

(i) The court supports the original judgment concerning "the suspension of illegal discharge of pollution."
(ii) In a change to the first trial judgment about the amount of damages for the agricultural and forestry products, now the defendant will pay the plaintiff a total of 684,178 yuan, not deducting past compensation payments.
(iii) The court dismisses the claim for payment for mental damage, the same as in the first trial.
(iv) In a change to the first trial judgment about the deadline for waste disposal treatment, the defendant is given an order to process the waste that was abandoned illegally within 1 year after this judgment becomes effective.

In comparison to the first trial judgment, three points on the second trial judgment could be found. Firstly, the request for the suspension of illegal discharge of pollution was admitted; secondly, a part of the request for the compensation for the economic damage was also admitted; and thirdly, the request for the compensation for mental damage was dismissed. Regarding the contents of the illegal discharge of pollution, the second trial judgment does not identify them, either.

On the other hand, the second trial judgment revised the first trial judgment on the following two points. First, the amount of compensation for damages that the second trial judgment admitted was slightly higher than first trial judgment, though it was still only approximately 6 % of the claimed amount. This is because the second trial judgment revised the first trial judgment, which deducted the compensation money paid to the local government by the defendant from the amount of compensation. In the first instance, this compensation money was paid to the local government and was not paid to the pollution victims. Therefore, this second trial judgment was modified to be within common sense. Second, the second trial judgment set a definite deadline for the removal or disposal of the illegally disposed of waste.

Meanwhile, the legal costs of the court of the first trial and the second trial were 90,000 yuan in total, and the total amount of the plaintiff's obligation was exempted. In addition, the court ordered that Rongping Chemical should pay the appraisal expenses for the trial, which were 100,000 yuan.

6.7 Execution Process of the Judgment

The second trial judgment of this case was given on November 16, 2005, as stated above. However, it was September 25, 2007, when the compensation for damages ordered by the judgment was carried out. Why was more than 1 year and 10 months needed to compensate the victims for the damages in this case? The process of this execution is helpful for thinking about the problem of the execution of judgments in China.

A month after the judgment in January 2006, the defendant, Rongping Chemical, brought the indemnification to the Pingnan County district court which was not concerned with this case at all. And the director of the Pingnan County district court kept refusing the payment of the indemnification to the representative of the plaintiffs.

The representative of the plaintiffs was at a loss and talked with the CLAPV. After the consultation, the vice director of the CLAPV proceeded to the location with five plaintiff representatives in July 2007 and negotiated with the director of the Pingnan County district court directly. Then the director stated, "I do not have the authority to settle this matter." Acquiring this answer from the director, the CLAPV vice director addressed a letter to the Chair of the Pingnan County Committee of the CPC on August 1, 2007.

This letter was transferred to the Pingnan County district court via Pingnan County People's Congress Standing Committee from the County Committee of the CPC, and a response to the letter was sent from the Pingnan County Court to the Pingnan County People's Congress Standing Committee on August 15 of the same year.

In this response to the letter, the direction from the County Committee of the CPC to the Pingnan County district court was written clearly. That is to say, the

County Committee of the CPC recognized the distribution of the indemnification could be a factor in social instability, so the County Committee of the CPC gave instructions with the following three steps to Pingnan County Court: firstly, suspend the indemnification temporarily; secondly, ask the representatives of the plaintiff to indicate the method for the distribution of the indemnification, so that all the members of the plaintiff group can agree to it; and finally, Pingnan County Court will confirm whether all the members of the plaintiff group agree to the distribution method of the indemnification.

In addition, according to the response to the letter of the Pingnan County Court, the Ningde City Intermediate People's Court supported the Pingnan County People's Court in following the instructions of the County Committee of the CPC, too.

The CLAPV vice director received this response to the letter and sent a protest to the Standing Committee of the County People's Congress. On the other hand, the members of the plaintiff group petitioned the provincial government and Provincial People's Congress. As a result of the actions mentioned above, the execution of the indemnification was accomplished on September 25, 2007.

With regard to the judiciary appraisal expense of 100,000 yuan that the plaintiff paid and the remitted legal costs, the counsel of the plaintiff and the plaintiff had to repeat similar inquiries and petitions, until it was paid by the Intermediate People's Court in July 2008.

6.8 Plaintiffs After the Trial

6.8.1 Clampdown on the Clinic of the Central Figure of the Plaintiffs by the Local Government

There was one clinic located approximately a hundred meters from the defendant's factory. The owner "Z" of this clinic was the only nonagricultural person among the five representatives of the plaintiff, and he was also the only person who could handle a computer, so it was he who communicated with the various media outlets and lawyers by e-mail. In other words, the reason why the pollution victims were able to contend at the trial was because "Z" could communicate with the media and legal specialists outside the village through e-mail.

On October 8, 2004, the Pingnan County Health and Medical Bureau clamped down on the clinic run by "Z" and ordered it to close for the reason that he conducted a medical practice without adequate permission. Furthermore, the Pingnan County Health and Medical Bureau ordered "Z" to pay a correctional fine ("fakuan" 罚款) of 5000 yuan.

Actually, "Z" started this clinic in 1983 and acquired a medical service occupation permit from the Ningde City government in 1993. We can find a lot of seals of the Pingnan County Health and Medical Bureau on this permit, and the seals of the Pingnan County Health and Medical Bureau were applied every year. These seals of the Pingnan County Health and Medical Bureau on the permit tell us that the permission was updated every year. Though the Health and Medical Bureau had allowed "Z" to serve in the medical occupation from 1993 through 2003 for 10 years, they refused to update the permit of "Z" suddenly and clamped down on his clinic in 2004. So, it is natural to think that the pollution suit and the closing order of the clinic are connected with each other.

"Z" expressed his disapproval with the decision of the Pingnan County Health and Medical Bureau and demanded an administrative appeal to the city Health and Medical Bureau on December 6, 2004. However, this appeal of "Z" was dismissed in January of the following year, and "Z" lost the administrative litigation regarding this issue, too. In October 2006, "Z" was suddenly detained for 30 h by public security, and during this period, the facilities of the clinic were forfeited forcibly by the local government.

6.8.2 Pressure on Social Groups

Some members of the plaintiffs of the suit formed an organization called "Green Home of Pingnan" (屏南绿色之家) in 2003. The purpose of the Green Home of Pingnan was to check the natural environment and farm products affected by the factory's pollution and to support the farmers who depend on the natural environment make their living.

Their activity was recognized as valuable by the Global Greengrants Fund (GGF) of the USA who support members of the private sector working on environmental protection in developing countries, and they received financial support of 3100 dollars in total from GGF. Then in 2007, their activity was also awarded a "SEE/TNC ecology prize," which was cosponsored by the Alxa SEE Ecological Association and the Nature Conservancy. In this way, the activities of the Green Home of Pingnan were highly valued globally as well as nationally in China.

However, they were arrested by local police officers, while they were on their way to distribute fliers in November 2006. Their overhead banner and handbills were forfeited, and the car that they rented was seized for 7 days. Furthermore, on a later date, September 27, 2007, when the plaintiffs received the indemnification of the pollution suit, they were called by the County Bureau of Civil Affairs. Two days later, the Green Home of Pingnan was punished as an unlawful group that moved into action without any report to the Bureau of Civil Affairs and was ordered to stop its activities.

6.9 Conclusion

By reviewing the process of the pollution dispute elaborated here, we can point out sympathetic attitudes of the local government and the local Committee of the CPC toward Rongping Chemical, in spite of the pollution damage that occurred around the company, and their impact on the local courts.

With regard to the decisions of the local government and the local Committee of the CPC, in the first place, both decisions expressed positive recognition about the invitation of Rongping Chemical to their county, saying it was supported by the CPC's leader at both the province level and city level, and then, they also admitted Rongping Chemical's contribution to the local finances and employment as an important factor for their economic and social development.

On the other hand, they acknowledged the fact that the pollution by the factory brought about some agricultural damage. In addition to this, they also admitted the necessity of a certain amount of compensation to the pollution victims, where we should notice that they also referred to their common goal of maintaining social stability and fostering mutual prosperity for both the factory and the village in that pollution area.

This feature of the local government or the local Committee of the CPC to use compensation for maintaining social stability in industrial pollution areas can be found in other districts, as has already been described in van Rooij et al. (2012), and what we found in this case is the impact of those decisions on the behavior of the local courts.

Both judgments of the first and second trials ordered the defendant, Rongping Chemical, to stop the illegal discharge of pollution as well as to pay compensation for the damages, but they did not state exactly what the defendant was to do to prevent pollution, nor did they indicate any criteria by which to measure the illegal discharge of pollution. As we elaborated above, the suggestions and opinions summarized by the Provincial EPA have already presented some concrete methods to guarantee the prevention of pollution, such as establishing a monitoring system or setting up a protection isolation zone, but the courts dismissed these suggestions by the EPA and kept the method and criteria vague.

Then, when the courts executed the judgments, they did not pay serious attention to the illegal discharge of pollution continuously causing damage to the villagers; on the contrary, the district court even suspended the execution of the compensation following the order by the local Committee of the CPC, and even the upper level court could not carry out their function to supervise the lower court to correct this unlawful execution process, giving way to the leadership of the local Committee of the CPC.

As a tentative conclusion, at least with regard to this case, we could say that the sympathetic attitudes toward the dominant industrial company demonstrated by the local Committee of the CPC in the process of the environmental litigation can have some impact on the local court, causing it to follow the decisions made by the local

Committee of the CPC, even while the EPA of the upper level government or the upper level court may not be able to have a substantial impact on the local court.

We need to elaborate on more cases to make a conclusion about the relationship between court judgments and decisions made by the local Committee of the CPC regarding environmental pollution, but we could say that if the courts have to follow the decisions of the local Committee of the CPC, then those decisions could be the actual limits for the courts to offer relief to pollution victims, just like those victims described in this paper.

References

Economy, Elizabeth (2004), The River Runs Black: The Environmental Challenge to China's Future, Cornell University Press

Stern, Rachel E. (2011), "From Dispute to Decision: Suing Polluters in China," China Quarterly 206, pp.294–312.

Stern, Rachel E. (2013), Environmental Litigation in China: A Study in Political Ambivalence, NY; Cambridge University Press, pp.50–70.

Wang, Alex (2007), "The Role of Law in Environmental Protection in China: Recent Developments," Vermont Journal of Environmental Law 8, pp.195–223.

van Rooij, Benjamin, Wainwright, Anna L., Wu, Yunmei, & Zhang, Yiyun (Amy) (2012), "The Compensation Trap: The Limits of Community-Based Pollution Regulation in China". Pace Int'l L Rev 29(3), pp.701–745.

Chapter 7
The Non-litigation Approach to Environmental Disputes in China: The Environmental Complaint Letters and Visits System – An Analysis

Shijun Zhang

Abstract Environmental dispute resolution generally includes settlements through litigation and non-litigation approaches. In light of the deficiencies in China's environmental legislation, the limited capabilities of courts in dealing with environmental disputes and the lack of public trust in courts, litigation has never been the main approach to resolving environmental disputes.

This article first looks back the history of the development of China's environmental complaint letters and visits system. Then, this article analyzes the necessity of environmental complaint letters and visits in environmental dispute resolution. This is not only because the environmental complaint letters and visits fit in with China's traditional culture and the current situation, either. More importantly, it is the most effective and low-cost way to resolve environmental disputes. But with the reform of the judicial system in China, especially since the environmental protection departments are not willing to involve themselves in environmental dispute resolution, the most recent environmental legislation tends to abolish the responsibilities of the environmental protection departments with regard to the resolution of environmental disputes. This article examines the defects of the current environmental complaint letters and visits system and suggests two major steps to improve it. The first is to set up a specialized agency for environmental dispute settlement. The second is to provide that environmental dispute settlement agency with the authority and the means to settle disputes.

Keywords Environmental complaint letters and visits • Environmental dispute settlement • Improvement

---

S. Zhang (✉)
Law School, Shandong University, Jinan, China
e-mail: zhangshijun@sdu.edu.cn

H. Kitagawa (ed.), *Environmental Policy and Governance in China*,
DOI 10.1007/978-4-431-56490-4_7

Abbreviations

ADR	Alternative dispute resolution
CGSS2006	Chinese General Social Survey
CPC	Communist Party of China
NPC	National People's Congress
PRC	People's Republic of China

7.1 An Overview of Environmental Complaint Letters and Visits in China

"Complaint letters and visits," shorten for "complaint letters and visits from the people," was originally adopted by the Communist Party of China (CPC). This system includes environmental complaint letters and visits and is a unique system with Chinese characteristics. It is an invention of the CPC, aiming to fulfill the ideological commitment of "serving the people" in such a way that encourages people to report issues, which will be subsequently investigated and solved by the government. That itself demonstrates the ideological effect of the party, which is to maintain close ties with the people and to achieve the goal of coming from the masses and going among the masses (Chen 2004). As a typical product of China, "complaint letters and visits" arose from a particular Chinese social and political environment. The system, historically derived from national leaders' instructions and working habits, was initially designed to solidify the relationship between the party and the people from top to bottom and promote better understanding and more support for the new regime (Ni 2010). With respect to legislation, Decisions on Handling People's Letters and People's Visit, a regulation promulgated by the Government Administration Council of the Central People's Government in 1951, is considered the starting point of China's modern system of complaint letters and visits.

In the early stages, complaint letters and visits were associated not with environmental disputes but mainly with political matters. The conclusion of one scholar's analysis is that the system of complaint letters and visits has gone through four primary stages (Chen 2006). The first stage, the preliminary formation stage of the system, was from the founding of the People's Republic of China (PRC) in 1957, during which the Government Administration Council of the Central People's Government promulgated Decisions on Handling People's Letters and People's Visit on June 7, 1951, and is generally considered to be the starting point of the formal establishment of the system (Chen 2006). During the second stage from 1957 to 1978, complaint letters and visits were mostly the result of mass mobilization, a considerable amount of which was related to political participation. The third stage was from the Third Plenary Session of the 11th Central Committee of the CPC in 1978 to the 12th National Congress of the CPC in 1982. The main task of the system in this stage was to solve a large number of historical issues and restore justice to falsely charged cases. Since 1982 to today, which is the fourth stage, the primary function of the complaint letters and visits system has transformed into dispute settlement and rights relief, with the political tasks in the third stage having been mostly completed. The last stage is a legalization process of the complaint letters and visits system, during which many associated laws and regulations have been published, such as Work Rules on Receiving Visits of the Complaint Letters and Visits, Department of the Supreme People's Court (最高人民法院信访处接待来访工作细则) (promulgated on June 20, 1980), Work Rules on Charging, Appeal, and Procuratorial Work of the Supreme People's Procuratorate (promulgated on December 10, 1986), and Regulation on Complaint Letters and Visits (promulgated by State Council on October 28, 1995). After the promulgation of Regulation on Complaint Letters and Visits, departments in the central government and many provincial and municipal governments also published regulations, complaint letters and visits working rules, and interim provisions successively.

Meanwhile, as Chinese society and the economy have been growing rapidly and public awareness of environmental protection has been enhanced, environmental problems have become an increasing concern, resulting in a substantial rise in environmental disputes and environmental complaint letters and visits. According to statistical data released by the China Environment Yearbook, the number of environmental complaint letters and visits started to surge in the mid-1990s. In 1995, the total national number of environmental complaint letters and visits was 58,678, while the number reached 608,245 in 2005. In order to improve the environmental complaint letters and visits system and to deal with environmental disputes, the State Environmental Protection Administration, based on the 1995 Regulation on Complaint Letters and Visits, enacted the Measures for Environmental Complaint Letters and Visits in 1997, which indicated that the Chinese government had begun to consider bringing environmental complaint letters and visits into the legalization process and that the Chinese environmental complaint letters and visits system was initially established.

With the deepening of reformation, the accelerating transformation of the economy and society, and the adjustment of the social interest structure, together with numerous new circumstances and new problems caused by those complaint letters and visits from the public, the State Council of PRC, facing an extremely difficult situation and the task of dealing with complaint letters and visits, published the revised Regulation on Complaint Letters and Visits in 2005. Subsequently, the State Environmental Protection Administration revised the 1997 Measures for Environmental Complaint Letters and Visits in 2006. The revisions of the two regulations promoted the further improvement of the Chinese complaint letters and visits system and the environmental complaint letters and visits system.

Based on more than 60 years of experience in complaint letters and visits and relevant laws since the founding of the PRC, the system of complaint letters and visits is a mechanism by which official offices that deal with complaint letters and visits take administrative or economic measures based on the policies of the CPC and of the state, provide consultations or mediation, and conduct political and ideological education to take care of various issues (Yang 2004). As a result, environmental complaint letters and visits is of particular significance in settling environmental disputes in Chinese society. One Chinese scholar has similarly asserted that in China, citizens' environmental participation is primarily conducted in the form of complaint letters and visits, proposals in the People's Congress and Political Consultative Conference, and administrative arbitration by the Environmental Protection Bureau and that environmental complaint letters and visits is the soundest system in the process of institutionalization of environmental participation by citizens (Zhu 2008).

7.2 Why Is Environmental Complaint Letters and Visits the Main Approach for Environmental Dispute Settlement in China?

As stated above, people in China seldom resort to lawsuits in the event of environmental disputes, the number accepted by courts being merely around 3 % of the total number (All China Environment Federation Website).[1] In the current process of legal modernization, the legal institutional arrangements of both litigation and non-litigation offer multiple choices for remedies and environmental dispute settlement, so why do people prefer complaint letters and visits, which is the non-litigation remedy? This inclination to choose complaint letters and visits instead of litigation has its historical and cultural reasons and relates to realistic predicaments of environmental litigation in China. Moreover, in contemporary China, complaint letters and visits has a special meaning when it comes to rights relief, and it has its own institutional value.

7.2.1 Environmental Complaint Letters and Visits Correspond with the Needs of the Chinese Traditional Legal Culture

The self-sufficient natural economy had been the dominant economic production mode of the Chinese agricultural society during the over 5000 years of feudal history of ancient China. By virtue of the characteristics of the natural economy, which were self-sufficiency, the underdeveloped technology in ancient society, and the lack of convenient and low-cost transportation, people confined themselves in a limited space, isolated themselves from each other, and kept insular and conservative views; thus, people in traditional society rarely went to the Kwana (the governmental office in Feudal China) to solve their disputes.

---

[1]http://www.acef.com.cn/site2/envlaw/news/38066.shtml

Confucius, the epitome of Chinese traditional culture, demonstrated this "lawsuit-detesting" sentiment: "When hearing cases, I, too, intend to keep them from happening again."[2] Therefore, it is safe to say that Confucius was very supportive of this non-lawsuit attitude, an attitude which is also evident in one of the classic works of Confucianism Chou, where the following is stated: "When things are not going well, it is better for a person to face it with forbearance. In case a lawsuit is involved, he shall maintain a peaceful mind, as the result will not be auspicious even if the problem is solved by the lawsuit. Therefore, in the event of a dispute, a prestigious man shall be relied upon to mediate instead of resorting to lawsuits."(Zhou Yi)[3] Against this background, disputes were usually resolved not by feudal officials but by clansman or other private organizations (i.e., guilds) (Liang 2002, Su 1996, and Zhang 2005).

7.2.2 The Realistic Predicament for Environmental Litigation in China

The judicial solution to environmental disputes is rather authoritative. However, environmental litigation is confronted by severe difficulties, mainly manifested as difficulties in presenting evidence, examination, and approval. Either the identification of causality between pollution and damage or the calculation of the amount of damages is extremely complex and requires professionals. At present, China has an enormous insufficiency in appraisal institutions for environmental pollution. As there are no uniform technique standards or methods, appraisal institutions sometimes come out with disparate or even opposite results for the same case, which puts the courts in a difficult position. Difficulties in case filing and high costs constitute the third reason for the environmental litigation dilemma. Victims of pollution cannot afford the expensive litigation costs and appraisal fees. Moreover, as polluting enterprises are sources of profit and taxes for local governments and the courts' expenditure comes from local revenue, the invisible local protectionism makes the situation of environmental litigation even harder. More than half of environmental lawsuits end up with the victims of pollution being the losing party, when they should have been compensated (Zhang 2003). In China, pollution victims tend to be a vulnerable group, especially people with low income. They are forced to give up on lawsuits, since they cannot afford the high litigation costs, resulting in complaint letters and visits becoming the primary solution for environmental disputes.

---

[2]Analects of Confucius·Yen-Yuan.

[3]Zhou Yi, Zhou Dynasty of Ancient China.

7.2.3 The Advantages of Environmental Complaint Letters and Visits in the Environmental Dispute Resolution Mechanism

China has a large number of environmental disputes, but the lawsuit-detesting culture combined with the predicament in practice pushes the public to stay away from courts and turn to a non-litigation approach, administrative organs in particular. Environmental complaint letters and visits, as one type of the complaint letters and visits system, has a comparative edge over lawsuits by providing a more convenient remedy approach for victims in environmental disputes. To begin with, environmental complaint letters and visits have relatively low costs. Official offices do not charge any fees but give complaint filers a certain amount in subsidies. Thus victims are more willing to choose environmental complaint letters and visits with the low-cost and flexible procedure rather than costly and time-consuming lawsuits with their complex procedure. Second, environmental complaint letters and visits have low evidence requirements. Complaint filers[4] need only to submit basic facts relevant to damages suffered from environmental pollution. Neither accurate evidence proving that enterprises have been emitting pollutants nor judicial authentication materials concerning personal or property damages are required. Specific conditions about the pollution caused by enterprises are investigated by official offices that deal with environmental complaint letters and visits, generally by environmental protection authorities at all levels through administrative processes. Damages are not necessarily based on judicially authenticated conclusions, but they may be investigated and confirmed during the subsequent dispute settlement process or may be presided over by environmental administrative organs and settled by the injuring party and the injured party through consultation. Compared with the rules concerning evidence in civil actions, environmental complaint letters and visits greatly alleviate the injured party's burden of proof and offer timely and efficient remedies. In addition, the environmental complaint letters and visits system could temporarily moderate the complaint filers' angst. It provides a channel for them to actively report environmental protection conditions and propose suggestions about environmental management or complain about environmental disputes and thereby pacify their mood and ease the situation. Unlike lawsuits, where cases are heard by judges, the reception staff at the official offices can sit down and have a conversation with complaint filers, during which they are listened to and get their questions answered. The existence of the environmental complaint letters and visits system keeps complaint filers from taking a

[4]According to the 2006 Measures for Environmental Complaint Letters and Visits enacted by the State Environmental Protection Administration, the complaint maker refers to citizens, legal persons, and other organizations that propose suggestions and opinions or make complaints about the conditions of environmental protection to environmental protection administrative departments at all levels in such forms as letter, email, fax, telephone, or visits.

more aggressive path to seeking relief and protecting their interests, thereby contributing to maintaining the stability of the social order.

Official offices that deal with environmental complaint letters and visits are generally set up inside the environmental administrative organs at all levels. They are entitled to organize, coordinate, and dispose of matters involving environmental complaint letters and visits on behalf of the corresponding environmental administrative organs,[5] which themselves have the responsibility of environmental supervision. According to the environmental management system established by the Environmental Protection Law of PRC, environmental protection authorities at all levels are the administrative departments in charge of the overall supervision of environmental protection in their administrative districts. As they are familiar with the local environmental conditions, environmental problems, and the status of enterprise pollution, they have a full understanding of environmental laws, regulations, and policies and are equipped with environmental technologies, environmental supervision methods, and environmental information, together with huge advantages in terms of fact-finding and evidence collection resulting from their authority and expertise. Environmental protection authorities at all levels can offset the deficiencies in the concerned parties' lack of ability to present evidence, and, with the help of environmental experts, they can accurately determine the facts, causes, responsibilities, and damages associated with environmental pollution, thus achieving a more reasonable result than would have been achieved through a trial. Environmental protection authorities are authorized to monitor environmental hazards; to deal with environmental pollution incidents and disputes; to exert, pursuant to related laws, various administrative measures against environmental tort-feasors, including, among others, on-site inspection, survey, sample detection, photograph, taping, evidence collection, supervision of the facilities for the prevention of pollution, and examination of the operation record; and to take compulsory measures against ongoing environmental infringement, which include but are not limited to administrative and legal means, such as warnings, fines, revocation of pollutant discharge permits, rectification within a prescribed time limit, and shutting down and moving. All of these powers enable the environmental protection authorities to directly deal with environmental pollution, reduce the cost of dispute settlement, and relieve the burden on the injured party, who will be compensated as soon as possible. Above all, environmental complaint letters and visits led by environmental protection authorities, in comparison with the courts' procedural

[5]Measures for Environmental Complaint Letters and Visits, article 8: Environmental protection administrative departments at the county level shall establish or designate an environmental complaint letters and visits office with full-time or part-time staff. Environmental protection administrative departments of provinces, autonomous regions, and districted cities shall establish environmental complaint letters and visits offices independently. The environmental complaint letters and visits offices of environmental protection administrative departments at all levels shall, on behalf of their corresponding departments, take charge of the work to organize, coordinate, deal with, and supervise complaint letters and visits and other related matters, and ensure the access to environmental complaint letters and visits.

functions, have evident advantages in terms of cost and expertise and have thus become the first choice for environmental dispute settlement.

7.3 The Challenges Confronting the Environmental Complaint Letters and Visits System

For a fairly long time, the environmental complaint letters and visits system has played a major role in settling environmental disputes. According to the provisions of the 2006 Measures for Environmental Complaint Letters and Visits, the main bodies of environmental complaint letters and visits activities are environmental protection administrative departments at all levels, and competent administrative departments of environmental protection at or above the provincial level must set up an independent environmental letters and visits work mechanism in light of the relevant provisions. As stipulated in article 15 of Measures for Environmental Complaint Letters and Visits, after environmental protection departments at all levels accept and hear environmental complaint letters and visits cases, they can organize relevant social groups, legal aid agencies, related professionals, and community volunteers to participate together in making use of counseling, education, negotiations, mediation, hearings, and other methods to legally deal with environmental issues reported by complaint filers, in a timely and reasonable manner. In accordance with article 41 of the 1989 Environmental Protection Act, administrative authorities usually, in the long-term practice of dealing with environmental complaint letters and visits, use administrative mediation and administrative processing methods to solve environmental disputes after they accept complaints from environmental victims. Before 1992, environmental administrative authorities, for the most part, settled environmental disputes by mediation, and if that failed, a failure was declared or an administrative decision was made. Practice has proven that it is difficult to reach a mediation agreement for the parties through mediation when it comes to serious environmental disputes. If administrative mediation fails, making administrative punishment decisions in a timely fashion (mainly in the form of administrative decisions or awards) is an effective process. With environmental departments that have a professional advantage in the driver's seat, dealing with environmental complaint letters and visits through, among other methods, administrative mediation has played a crucial role in environmental dispute resolution. According to the data of the Chinese General Social Survey in 2006 (CGSS2006) conducted by the Department of Sociology of Renmin University of China and the Department of Social Sciences of the Hong Kong University of Science and Technology, when facing environmental disputes, most people (59.4 %) prefer to choose non-institutionalized methods, especially in the form of private relief to directly negotiate with polluters (40.5 %). As indicated in Table 7.1, when resorting to institutionalized methods, most people (39.5 %) hope to solve the disputes through the government, which is mainly in the form of

Table 7.1 Survey of action modes

Mechanism of dispute settlement		Action mode	Sample number	Proportion (%)
Institutionalized dispute solution	Litigation	Filing a suit	2	1.1
	Administrative punishment	Turn to government departments	73	39.5
Non-institutionalized dispute solution		Negotiate with polluters directly	75	40.5
		Reflect to the media	7	3.8
		Others	28	15.1

Source: Tian, Meng, Theoretical Analysis and Practical View on the Environmental Dispute Resolution Mechanism [J], Law Review, 2015(2), 176

complaint letters and visits and complaints. Only 1.1 % of the people hope to settle such disputes through litigation.

However, in cases of involvement of administrative measures when dealing with environmental letters and visits and environmental disputes, if any party is not satisfied with the administrative punishment decision and files a lawsuit, the environmental protection departments may become the defendant of the lawsuit. Therefore, environmental protection departments are not willing to take the responsibility of dealing with environmental disputes or to intervene in disputes between polluters and victims to avoid trouble. In recent years, during the process of the revisions of environmental laws, the Ministry of Environmental Protection has used its advantageous position in drafting amendments to gradually modify and derogate its responsibility to deal with environmental disputes through administrative methods. The responsibilities of the environmental protection departments in environmental dispute resolution were stipulated in article 41 of the 1989 Environmental Protection Act, which was in effect for a long time in China: "A dispute over the liability to make compensation or the amount of compensation may, at the request of the parties, be settled by the competent department of environmental protection administration or another department invested by law with power to conduct environmental supervision and management. If a party refuses to accept the decision on the settlement, it may bring a suit before a people's court. The party may also directly bring a suit before the people's court." The ways for environmental protection departments to deal with environmental disputes mainly include administrative mediation and administrative adjudication. But due to the fact that the use of administrative adjudication to deal with environmental disputes may lead the environmental protection department to become the defendant in the lawsuit, environmental protection departments at all levels have tried to avoid the use of administrative decisions. Efforts by environmental protection departments to perform fewer duties in environmental disputes by trying to influence legislation began on January 31, 1992, in the environment protection leading group of the state council, which was drafted by the National People's Congress (NPC) Law Committee Office. The main content of the approval is as follows: "When the parties concerned refuse to accept the compensation decision and bring a suit in a people's court, they shall not take the environmental protection departments as defendant of the administrative proceeding but shall file a civil lawsuit against the original pollution compensation disputes. At the same time, the people's court shall not accept and trial the disputes as administrative cases." Though the approval does not conform to the regulations of legal interpretation in the Law on Legislation of the PRC, the legal interpretation is considered to be that environmental departments be exempt from administrative means when dealing with environmental disputes. In fact, this approval has affected the written responsibilities of environmental protection departments in dealing with environmental disputes. The Law of the People's Republic of China on Prevention and Control of Environmental Pollution by Solid Waste enacted in 1995, the Law of the People's Republic of China on the Prevention and Control of Atmospheric Pollution revised in 2000, and the Water Pollution Prevention and Control Law revised in 2008 all stipulate, "Disputes in liability or amount of compensation can be mediated by a competent administrative department for environmental protection on the basis of the request of the parties." In the latest revision of the Environmental Protection Law, effective in 2015, the responsibility of the environmental protection departments in environmental dispute resolution is completely abolished. With regard to dispute resolution, Article 64 of the new Environmental Protection Law stipulates, "Those who cause damages due to environmental pollution and ecological destruction shall bear tort liability in accordance with provisions of Tort Liability Law of the People's Republic of China."

It has been difficult for environmental complaint letters and visits to play their due role in environmental dispute resolution considering that the main responsible departments are gradually administrative measures such as mediation and adjudication as well as various defects in the environmental complaint letters and visits system itself.

7.3.1 The Complaint Letters and Visits Agency Is a Single Mandate with Restricted Privileges

According to the Regulation on Complaint Letters and Visits, the complaint reporting handling office fulfill the following duties: accepting, assigning, and transferring the matters as described in the complaint letters and visits, handling the matters as described in the complaint letters and visits as assigned by the people's governments at and above the same level, coordinating the handling of important complaint letters and visits, and supervising and checking the handling of matters as described in the complaint letters and visits. As it can be seen from the above provisions, complaint letters and visits agencies are confined to procedural matters like acceptance, assignation, transference, coordination, and supervision. In short, even if the complaint letters and visits agency requires further verification of the situation, it merely "may" investigate other organizations and persons and is only entitled, regarding specific matters concerning complaint letters and visits, substantive issues that are not explicitly stipulated in the Regulation on Complaint Letters and Visits with "the three suggestions." The complaint letters and visits agencies have no independent authority to solve environmental dispute problems. Therefore, it is rather cumbersome to completely resolve the problem of people's complaint letters and visits from a legislative perspective.

7.3.2 The Lack of Operating Specifications

Although the system of complaint letters and visits has been established as a legal system, it lacks clear internal operating specifications. According to the Regulation on Complaint Letters and Visits, the work of complaint letters and visits agencies mainly consists of reporting, transferring, delivering, and informing complaint filers of other solutions, while the only matter directly handled by the agencies is the absence of specific process requirements. In practice, complaint letters and visits agencies generally do not directly deal with cases. Most act as a "transit station" of diversion. The higher the level of the agency, the more apparent this phenomenon appears. Complaint letters and visits cases are submitted, transferred, and sent to relevant offices by complaint letters and visits agencies. Sometimes such cases will be sent back to the office where the complaint filer works, which contributes to retaliation against the complaint filer. Meanwhile, although the Regulation on Complaint Letters and Visits sets a time limit for handling concerned matters, it does not provide a notification system proving relevant information concerning the results.

7.4 Key Points for Improvement of the Environmental Complaint Letters and Visits System

Presently, China's complaint letters and visits system is facing a predicament, and there is controversy brewing over whether or not the system should be retained. This situation is mainly a result of the following on the following reasons: "After the complaint letters and visits system was established, it once played an effective and subtle role in easing social contradictions and safeguarding the stability and unity of social life as a whole. However, along with the rapid development of China's economy, conflicts of interest have further intensified; with the system in transition, a lack of rules is leading to a phenomenon of corruption and social injustice. From all these problems, it is evident that the function of the complaint letters and visits system needs to be changed. The original function of contact with the masses and reflecting public opinion is in a trend of weakening, but the function of rights relief is rising to a more important position, and the complaint letters and visits system has become the main means for a considerable portion of the public to fight the government. The trend of making use of the complaint letters and visits system for political organization and involvement in foreign affairs will become a great threat to society and the government. The system is doubted by all circles of society. As a result, to retain the system of complaint letters and visits or not has become a focus of the current debate."[6] The environmental complaint letters and visits system faces predicaments as well. Analyzed externally, the direction of China's judicial reform is gradually turning toward the advocacy of the rule of law. Thus, when environmental disputes arise, it is more attractive to solve them through court proceedings. Especially when the environmental protection departments do not want to take responsibility for solving environmental disputes, it is more reasonable to turn to the court for victims of environmental pollution. From the point of view of environmental complaint letters and visits system itself, there are also plenty of disadvantages to the settlement of disputes.

These two aspects are causing great harm to the handling of environmental disputes by the environmental complaint letters and visits system. Advocating the rule of law does not mean that the environmental disputes must be resolved through legal channels. The court does not have any advantages in dealing with environmental disputes; on the contrary, it faces many difficulties in dealing with the problems. In fact, Chinese courts have been encumbered with numerous cases. Currently, the method of court-entrusted mediation[7] is being experimented within dispute settlement in order to reduce the appeal of the court.[8] The problems with the environmental complaint letters and visits system itself can be solved through compensating for the institutional drawbacks to eventually achieve a better solution to environmental disputes. China is still in a stage of industrialization. Although the growth of the economy is slowing down, it remains one of the countries with the fastest economic growth in the world. And conflicts caused by environmental disputes will exist in China for a long period of time. Therefore, it is necessary to review the drawbacks in the current environmental dispute processing mechanism carefully, improve the environmental complaint letters and visits system, and find an effective way to resolve disputes. Japan, South Korea, Taiwan of China, and other countries and regions whose economies have developed a lot in the East Asia region have experienced a period of sharp pollution conflicts caused by rapid economic development. These countries and regions have a similar culture and social background to China, and they are all densely populated areas, so it is worthwhile

[6]Huang Songjie. On the path of the reform of the system of letters and visits under the background of social transformation [J/OL]. Forum of Rule of Law, http://bbs.chinacourt.org/index.php?showtopic=390398

[7]Court-entrusted mediation means that the court gives the civil case out before or after the case is accepted to a specific organization or individual to host the mediation, also known as "entrusted mediation" or "sendout mediation." See also Liu Jiaxing, Pan Jianfeng, editor in chief: "Civil Procedure Law Course," Peking University Press, 2008, p 178.

[8]Provisions of the Supreme People's Court about Several Issues Concerning the Civil Mediation Work of the People's Court.

for China to learn about their environmental dispute resolution mechanisms from them. Considering the main problems of China's environmental complaint letters and visits system itself and the environmental dispute solution system of these countries and regions, the emphasis should be laid on setting up a specific dispute processing mechanism and giving rights and authority to these mechanisms to settle disputes.

7.4.1 Setting Up Specific Organizations for Environmental Dispute Processing

7.4.1.1 Environmental Public Nuisance Dispute Coordination Commission System in Japan

In order to govern the pollution and protect the environment, the government of Japan issued the Act on the Settlement of Environmental Pollution Disputes in 1970, in which the provisions of the administrative system of environmental disputes were formulated. The Act established the Environmental Dispute Coordination Commission and the system of public nuisance affliction talk. These two are similar to some extent to the environmental complaint letters and visits system in China.

In this administration system, the central government set up the Environmental Dispute Coordination Commission as a higher leading body and set up the Prefectural Pollution Review Board as a specific agency to deal with pollution complaints in every prefectural county and city, town, and township (including special districts) of Japan. There are public nuisance mediators in the Environmental Dispute Coordination Commission. The Environmental Dispute Coordination Commission must be composed of a chairman and six committee members. They are appointed by the Japanese prime minister after approval by Congress. In order to ensure the neutrality and independence of Environmental Dispute Coordination Commission, the role and function for the chairman and committee members within their term of office is protected by law and is not easily changed due to external interference.

7.4.1.2 Environmental Dispute Resolution Commission in South Korea

South Korea promulgated the Environmental Disputes Adjustment Act (also known as the Environmental Pollution Victims Dispute Processing Law) in 1997, which entrenched the administrative settlement mechanism of environmental disputes. Environmental Dispute Adjustment Committee, which is established according to the Act, is the main body of the administrative settlement mechanism of environmental disputes. The Commission is composed of two levels—the central level and the local level. The central level is appointed by the President of South Korea, and the local level is appointed by regional and district administrations.

7.4.1.3 Environmental Dispute Resolution Agency in Taiwan

In 1992, Taiwan promulgated the Public Nuisance Disputes Mediation Act. The Act provides a public nuisance dispute mediation committee and directly controlled municipalities and governments of counties (cities) that must set up full-time personnel to handle the following matters: (1) dealing with pollution complaints; (2) handing necessary investigations, guidance, and suggestions for pollution complaint processing; and (3) directing those who file complaints to ask for mediation or arbitration in accordance with the procedures of the law.

According to the Measures for Environmental Complaint Letters and Visits of China,[9] the department responsible for environmental letters and visits work is not an independent agency. It is an inner branch of the administrative departments of environmental protection and has no post or power to deal with environmental disputes independently. In China's administrative management system, an agency without independent human resources, financial resources, or material resources has no independent capacity to manage its responsibilities. Thus, its capacity and final effect is also greatly reduced in environmental disputes processing. In improving the institutional settings of the environmental letters and visits mechanism, the Environmental Complaint Letters and Visits Processing Center or Environmental Dispute Coordination Commission or other similar institutions can be considered set in the Ministry of Environmental Protection, and environmental administrative organs at all levels set up the corresponding Environmental Complaint Letters and Visits Processing Center. Environmental administrative organs at all levels must be responsible for environmental disputes within their jurisdiction of the area under the principle of territorial jurisdiction, and environmental disputes crossing administrative areas should be solved by level jurisdiction. The Environmental Complaint Letters and Visits Processing Center should consist of staff working in the environmental administrative organs, judges in the people's court, environment experts, legal experts, and people in environmental protection organizations and scientific and scholarly circles. Moreover, representatives of area businesses and the masses and the neighborhood committee staff are to be involved in the center.

7.4.2 The Power and Method of Environmental Disputes Litigation: The Global Experiences

In order to fully be effective in disputes, not only specialized agencies for environmental disputes but also the corresponding power is needed.

[9]According to article 8 of Measures for Environmental Complaint Letters and Visits of China, only the provincial environmental protection departments and the larger city divided into districts must set up a separate environmental letters and visits work agency.

Japan's "Public Nuisance Disputes Mediation Act" provides for administrative handling of environmental disputes, including non-litigation solution, mediation, arbitration and alternative dispute rulings, and other solutions like administrative litigation lawsuit solutions. Japan's pollution adjustment committee is responsible for handling environmental dispute complaints from the public, including ruling the causal relation of environmental disputes and the recognition of liability.

Public nuisance dispute mediation committee in accordance with South Korea's Environmental Disputes Adjustment Act serves as an environmental administrative dispute settlement body. The committee, as a quasi-judicial body, has the power to make unilateral rulings. Once the parties accept mediation for environmental disputes at the Commission's recommendations, the environmental administrative system is the only effective way to solve environmental disputes. South Korea's environmental dispute administrative solutions consist of three methods: pre-conciliation, mediation, and arbitration. The first two do not produce a legal effect unless the parties voluntarily submit to a conciliation committee; the latter has the force of law, and the parties accept an agreement that they will no longer bring about civil action. With early pre-mediation-conciliation procedures, such as when the parties resolve a dispute in the early mediation program, conciliation proceedings should no longer be conducted. Environmental pollution victims can apply directly for an arbitration procedure, but the Commission found that when mediation is more conducive to the settlement of disputes, arbitration proceedings may be terminated and conciliation proceedings begin. An application can be filed for civil proceedings for a resolution when mediation and arbitration cannot resolve disputes.

According to the relevant provisions of Taiwan's "Public Nuisance Disputes Mediation Act," the committee dealing with public nuisance dispute mediation mainly concentrates on municipalities, counties, and the "EPA" ruling committee. These two bodies, "as the trial court level, are responsible for applying public nuisance dispute mediation and accepting the award" (Cheng 2003). Obviously, mediation and adjudication of environmental disputes has a statutory approach in Taiwan. Mediation means that a mediation committee should consult the parties with a dispute to mediate, to persuade the parties that the two sides can be led to reach a settlement agreement, or, at the request of the parties concerned, to investigate the relevant facts of the dispute and accordingly offer a mediation proposal, and if the parties do not file an opposition within a specified time, then this program is established. Implemented after the establishment of mediation and mediation by court review, it has the same effect as a court decision, and the parties shall not sue for the same matter at the court again, and the parties may also request that the court enforce the mediation.

According to China's current Measures for Environmental Complaint Letters and Visits, the duties and powers of the environmental petition mechanism are limited to receiving and forwarding and are assigned to coordinate the environmental petitions process, its function being limited to procedural powers with no independent functions or powers to handle environmental disputes, the result of which is that in dealing with environmental disputes petitions, it has no legally binding mechanism.[10] Thus, the petition agencies dealing with environmental disputes in effect are naturally greatly reduced. In Japan, South Korea, China, Taiwan, and other environmental dispute mechanisms, they are able to fulfill their dispute settlement functions. The reason for this is that these institutions have a variety of means, including mediation, arbitration, adjudication, and awards. South Korea's environmental dispute resolution board has the authority to make unilateral decisions, which also have a legal binding force. Thus, they are more efficient in handling environmental disputes.

On the perfection of China's environmental petition mechanism of responsibility and authority, what we can do to explicitly deal with environmental disputes are mediation, adjudication, and administrative arbitration (including the existing administrative rulings, decisions, etc.). First, the administrative mediation can take a variety of forms, like mediation, negotiation, etc., and after complying with statutory procedures, mediation documents could have a legal effect (the mediation documents could be as valid as the reconciliation agreement under the adjudication). If the mediation fails, the complaint-handling institution has the right to exercise the power of administrative adjudication. Second, with administrative rulings, this procedure is started by the application of the administrative party, which is a kind of quasi-judicial procedure. In terms of identification of legal facts, application of the law, and solving disputes, it differs from mediation and arbitration inessential ways. As far as use of expertise to identify facts is concerned, it is different from judicial means, and it can be said to be an administrative trial. The main contents of the ruling include the determination of the cause of damage and the liability of tort, the remedy (mainly the amount of compensation), and the production of administrative adjudication. The administrative adjudication should be legally binding.

7.4.3 Chinese Local Legislation Breakthrough in Terms of Environmental Petition

In fact, as early as the 1990s, many provinces in China had established rules on the development of effective environmental petitions and dispute regulations through local legislation.[11] The 2014 Sichuan Provincial Environmental Protection Bureau issued the Sichuan Environmental Dispute Mediation Level Management Rule taking full advantage of existing environmental complaint-handling mechanisms. Very detailed, it clearly provides for the establishment of an environmental disputes body to exercise the functions of the bureau, and it is worth using for reference.

[10]See Environmental Petition Approach article 10.

[11]Such as "Henan Environmental Pollution Disputes Interim Measures"(1991), "Hubei Environmental Dispute Mediation Rule" (1992), "Qinghai Environment Damage Compensation Dispute Mediation Rule" (1992), and Shandong Provincial Environmental Pollution Disputes Rule" (1994)

First, institutions should be established to ensure professional mediation of environmental disputes. On the basis of an external linkage mechanism, which includes environmental protection, the SASAC, courts, prosecutors, and other departments involved in mediation, an environmental dispute mediation committee should be set up in the environmental protection system and be divided into three classes: provincial, city, and county. The committee should consist of department heads of relevant businesses and be responsible for the organization and coordination of environmental dispute resolution within its purview. The professional mediation committee should have a mediation room. Those who have the authority should deploy people who are professional and skilled mediators to be responsible for the investigation in order to study and solve environmental disputes. Concerning the establishment of township (neighborhood) organizations, we should take those into consideration and make full use of the existing township (neighborhood) people's mediation and judicial mediation bodies, setting the specific person to be responsible for the mediation of environmental disputes. In addition, on the team construction issue, environmental protection departments at all levels should broaden their channels, absorbing the relevant departments, experts, lawyers, etc., to enrich the mediation team. On the job security issue, monetary requirements for dispute resolution should be incorporated to the environmental budget funds.

Second, hold a meeting to ensure that environmental disputes risk assessment is standardized.

Environmental dispute-developed risk assessment criteria must be established, which requires the city (prefecture), county (city, district), and township (neighborhood) to assess the risk within 3 days according to the length of time of the environmental dispute, the number of people involved, and the degree of difficulty of the settlement. At the same time, an environmental dispute rating on a scale from grade III to grade I must be determined. Events that may lead to significant environmental disputes require timely assessment and in 1 day to clear a risk level and responsibility units.

Third, hierarchical management should be implemented in a concrete way to ensure that there is a clear division of responsibilities.

In the Sichuan Provincial Environmental Dispute Mediation Level Management Rule, the responsibilities associated with environmental dispute mediation and classification management are divided. The evaluation grade III (the same as mentioned in the part "Firstly") environmental disputes are organized by the environmental protection departments at the county level, and they are resolved by the villages and towns (street); evaluations for grade II environmental disputes are resolved by the county (city, district) environmental protection departments who coordinate responsibilities with the units that have the relevant responsibilities; evaluations for environmental disputes at grade I are based on the municipal environmental protection departments who coordinate with the units that have the responsibility to seek resolution. For the counties (cities, districts), which include multiple departments and the municipal level grade I difficult environmental disputes, the municipal environmental protection department must report to the people's government or mediation center. There must be also a clear linkage to the leader and cooperative units that can resolve this problem. At the same time, we also implement a classification of dynamic management of environmental disputes, and when environmental disputes have been assessed as having escalated requirements, they must be reassessed, and the risk level and the responsibility of the units also are clearly defined.

References

Chen, Bofeng, Vexatious, Suits, Complaint Letters and Visits and Chinese Legal Tradition—Issues Concerning Vexatious Suits During the Transforming Period of Law (缠讼、信访与新中国法律传统——法律转型时期的缠讼问题) [J], Peking University Law Journal, 2004(2).

Ni, Yujie, The Historical Review and Current Situation of Complaint Letters and Visits in China (我国信访制度的历史回顾与现状审视), Chinese Administration Management, 2010 (11).

Chen, Jianbing, Research on the Complaint Letters and Visits System (信访制度研究)[D], Shanghai: Thesis of Master Degree of East China University of Politics and Law, 2006.

Yang, Yanwei, Jurisprudential Analysis of the Complaint Letters and Visits System(信访制度的法学透析)[D], Beijing: Thesis of Master Degree of China University of Political Science and Law, 2004.

Zhu, Xufeng, The Role of Environmental Citizen Society in Environmental Governance(环境公民社会在环境治理中的作用) [J], China Social Sciences Documentation, 2008 (10).

Liang, Zhiping, Seeking the Harmony in Natural Order(寻求自然秩序中的和谐), ISBN 2002.

Su, Li, Rule of Law and Its Local Resources(法治及其本土资源), ISBN, 1996.

Zhang, Jinfan, The Tradition and Modern Transition of Chinese Law (the Second Edition)(中国法律的传统与近代转型》(第二版)), Law Press China, 2005.

Zhang, Shijun, The Analysis of the Administrative Resolutions of Environmental Disputes(论环境纠纷的行政处理), The Rule of Law Forum, Nov., 2003.

Liu, Jiaxing, Pan Jianfeng, "Civil Procedure Law Course"(民事诉讼法学教程), Peking University press, 2008.

Chen, Chi Yang: "Environmental Law Introduction"(环境法总论), China University of Political Science Press, 2003.

How to Safeguard One's Right while Facing the Pollution Accident, How Far can the Environmental Court Go, available at: http://www.acef.com.cn/site2/envlaw/news/38066.shtml.

Chapter 8
Review of the Legislation on Public Participation in EIA in China: From Disorder to Normalization

Shekun Wang

Abstract China began to implement environment impact assessment (hereinafter referred to as EIA) in the late 1970s; however, public participation had not been introduced into EIA procedure until 10 years later. Accordingly, since the late 1990s, public participation in EIA in China started from scratch and has gone through the development process from disorder to normalization. China's early legislation on public participation in EIA was too vague and impractical, which led to the disorder of practice in public participation in EIA. In 2006, China enacted special measures, making public participation in EIA achieve a preliminary legalization. Deep-seated problems in the practice of public participation in EIA in recent years, such as the lack of validity and authenticity of public participation, have presented new challenges for the legislation on public participation in EIA. China has begun to revise the Interim Measures, intending to resolve core issues, such as the scope of application for public participation in EIA, the criteria for how "fully" the public opinions are solicited, and the effect of public opinions, so as to improve the authenticity and validity of public participation.

Keywords Environment impact assessment • EIA • Public participation • Legislations • Public opinions • Information disclosure • Participation approaches

Abbreviations

EIA	Environment impact assessment
EIR	Environmental impact report
EPL	Environmental Protection Law
LEIA	Law on environment impact assessment
MEP	Ministry of Environmental Protection
NEPA	National Environmental Policy Act

S. Wang (✉)
Law school at Peking University, Beijing, China
e-mail: wsk@pku.edu.cn

H. Kitagawa (ed.), *Environmental Policy and Governance in China*,
DOI 10.1007/978-4-431-56490-4_8

SC State Council
SEPA State Environmental Protection Administration
SETC State Economic and Trade Commission
SPC State Planning Commission

8.1 Introduction

Since the National Environmental Policy Act (hereinafter referred to as NEPA) of the United States stipulated an environmental impact assessment (hereinafter referred to as EIA) for the first time in 1969, EIA was quickly adopted as an effective tool for environmental management in environmental legislation all over the world. From the viewpoint of the historical background of EIA, the function of EIA can be summarized in the following two aspects: scientific decision-making and democratic decision-making. Correspondingly, the alternatives in designing EIA and public participation are regarded as the heart and soul of EIA (Wang 2006b: 226).

In the late 1970s, on the basis of the experience of Western countries who were pioneers in conducting EIAs and China's actual situations, China established an EIA with Chinese characteristics. Different from the practice in Western countries of emphasizing the process of public participation in EIA, China gave a higher priority to the scientific decision-making aspect of designing EIA, the core of which was the closed-end administrative process of environmental administrative organs including reviewing and approving EIA documents (Wang 2006a: 38). Correspondingly, there were no procedural requirements for public participation in China's EIA legislation in the beginning.

However, with the promotion of democratic awareness and environmental consciousness, this closed-end design of the EIA process not only eventually revealed the shortcomings of the lack of public opinion basis for EIA decisions but also caused a lot of disputes and arguments. With the deepening of China's reform and opening up, China began to introduce public participation into EIA in loan projects by the World Bank, the Asian Development Bank, and other international financial institutions. In the 1990s, China's legislation on public participation in EIA began to appear. It started from scratch and has since gone through the development process from disorder to normalization. In the following years, the evolutionary process of China's legislation regarding public participation in EIA will be examined, the existing problems will be revealed, and the direction for revising China's legislation regarding public participation in EIA will be explored.

8.2 The Evolution of China's Legislation on Public Participation in EIA

8.2.1 Beginning of China's Legislation on Public Participation in EIA

After China's first National Conference on Environmental Protection, which was held in 1973, EIA was introduced into China and established as an environmental legal institution in Environmental Protection Law (for trial implementation) (环境保护法试行) in 1979 for the first time. In 1981, the former State Planning Commission (hereinafter referred to as SPC), the former State Economic and Trade Commission (hereinafter referred to as SETC), the former State Construction Commission, and the former Leading Group for Environmental Protection in the State Council (hereinafter referred to as SC) issued the Measures for the Administration of Environmental Protection of Fundamental Construction Projects (基本建设项目环境保护管理办法), stipulating the scope, procedures, methods, costs, approval, and other specific regulations for the EIA for fundamental construction projects, thus making EIA more tractable. In 1986, the former Environmental Protection Commission of the SC, the former SPC, and the former SETC jointly issued the Measures for the Administration of Environmental Protection of Construction Projects (建设项目环境保护管理办法), expanding the scope of EIA from fundamental construction projects to technical transformation and regional development projects. Article 13 of the Environmental Protection Law (hereinafter referred to as EPL), enacted in 1989, provided general stipulations for EIA again. Based on the contents of the aforementioned laws and regulations, the design of EIA at that time did not have any procedural requirements for public participation. This legislative choice was closely related to the backdrop of China's planned economy and highly centralized political system at that time.

After entering the 1990s, with the development of aid programs from international organizations in China, the issue of public participation in EIA was incorporated into the topics being discussed. In 1991, China implemented a training program on EIA granted by the ADB, and the issue of public participation in EIA was discussed in China for the first time. Subsequently, public participation has become a hot issue in the field of EIA in China (Li and Li 1998).

In 1993, the former SPC, the State Environmental Protection Bureau, the Ministry of Finance, and the People's Bank of China jointly issued the Notice on Strengthening the Administration of EIA for Construction Projects Loaned by International Financial Organizations (关于加强国际金融组织贷款建设项目环境影响评价管理工作的通知), in which it was expressed for the first time that "As an important component of EIA, public participation should be detailed in a special chapter set up in the report," and the subject and basic approach to public participation in EIA was stipulated.

Up until this point, driven by external forces, public participation had become a new area and focus of EIA in China (Li 2002), and relevant legislation on public participation in EIA began to appear. For example, the Law on Prevention and Control of Water Pollution (水污染防治法) revised in 1996, the Law on Prevention and Control of Pollution from Environmental Noise (环境噪声污染防治法) enacted in 1996, and the Regulations on the Administration of Environmental Protection of Construction Projects (建设项目环境保护管理条例) enacted in 1998 all required that "the opinions from related units and residents at the location of construction projects should be solicited in EIA." However, the aforementioned legislation only provided for the principle of public participation, while specific problems such as the scope of the public and the means by which the public could participate were not involved.

8.2.2 The High-Speed Development of China's Legislation on Public Participation in EIA

Because of the importance of EIA, China enacted the Law on EIA (hereinafter referred to as LEIA) (环境影响评价法) in 2002, which was a milestone in the development of EIA in China. Meanwhile, the LEIA laid the legal foundation for the development of public participation in EIA. First, Article 5 in the general rules of the LEIA declared the State's attitude toward public participation in EIA, that is, "The State encourages relevant units, specialists, and the public to participate in EIA in an appropriate manner." Second, Articles 11 and 21 of the LEIA defined the time to participate in EIA, the scope of the public, the approach to participation, and supporting measures separately in public participation in EIA for both planning and construction projects, which formed the institutional framework for public participation in EIA in China.

There is no doubt that the LEIA has progressed more significantly forward in public participation in EIA than the former legislation, but vague items and weak operability in the LEIA still lead to problems in public participation in EIA. The Yuanmingyuan anti-seepage engineering incident (Focusing on Yuanmingyuan... 2015) is a perfect example.

In April 2005, the former State Environmental Protection Administration (hereinafter referred to as SEPA) held a public hearing concerning the environmental impact of the Yuanmingyuan anti-seepage engineering incident taking advice from experts, social organizations, the public, and relevant administrative organs. It was the first hearing held by a national central administration and could be seen as a significant model in the history of public participation in EIA. However, this hearing fully demonstrated the problems with public participation in EIA in China at the same time. After the hearing, the SEPA issued a notice called "On Forcing the Yuanmingyuan Management Office to Refile an EIA Report in Due Time" (关于责令限期补办圆明园环境综合整治工程环境影响评价报批手续的通知), and after that, no public participation can be found in either the drawing up or examining and approving EIA documents. The hearing was held where no legal requirements for public participation existed, but no hearings were held at all where required by law. It could be said that the hearing for the Yuanmingyuan anti-seepage engineering incident greatly promoted public participation in EIA in China, through which the public began to realize that public participation should be an indispensable part of an EIA. However, this hearing contributed nothing to the institutional construction of public participation in EIA; on the contrary, the roughness and shortcomings of the legislation on public participation in EIA were exposed.

After the Yuanmingyuan anti-seepage engineering incident, the level of social concern about public participation in EIA increased significantly, and public participation became a hotspot and difficult part of EIA (Wu et al. 2012). Since the provisions for public participation in the LEIA were too vague and could not fulfill the practical needs, in 2006, the former SEPA issued the Interim Measures on Public Participation in EIA (hereinafter referred to as Interim Measures) (环境影响评价公众参与暂行办法), in which the time, the approach, the scope, and the supporting measures for the public to participate were stipulated in detail. Up until this point, the LEIA and the Interim Measures had formed the main part and legal basis of China's legislation on public participation in EIA. The high-speed development stage ended, and then the stage of implementation and perfection began.

8.3 Problems with China's Legislation on Public Participation in EIA

During the 10 years since the Interim Measures were issued, public participation in EIA in China has made great progress, and the role of the public has become more and more important. For instance, since the Yuanmingyuan anti-seepage engineering incident, in the EIA for the Shanghai maglev train project, the Xiamen PX project, the Beijing Liulitun waste-to-energy project, the Jinshajiang hydroelectric development project, and other large-scaled construction projects, the public all played an active role and elevated the effectiveness of the EIA (Jin 2012). However, taking the conflict of the EIA in the project of the waste incineration plant in west Qinhuangdao (Yang 2013) as an example, due to those problems in legislation on public participation in EIA, the phenomena of distortion and invalidation in the public participation exist widely in EIA practice. These problems will be analyzed below on the basis of concluding the main contents of China's legislation on public participation in EIA.

8.3.1 Objects of Public Participation in EIA

Depending on the extent of environmental impact, there are two types of EIA in active laws and regulations that have major influence on the environment and thus need public participation: one is the EIA for construction projects with an environmental impact report (hereinafter referred to as an EIR) and the other is the EIA for special planning with an EIR. This partition in law has caused a lot of conflicts in EIA, since the extent of environmental impact has no inevitable relevance for the extent of social concern. The EIA for comprehensive planning and the EIA with an environmental impact sheet or environmental impact registration form, where no public participation is needed according to the legal provisions, are always of great social concern, especially from the residents at the site of the projects. However, the constant lack of legal channels for participation often leads to mass disturbances (Zhu 2015).

8.3.2 The Scope of "The Public"

China's legislation on EIA defines "the public" officially as "relevant units, experts, and the public." The criterion of "relevant" is based on whether the subjects are influenced, which include potential beneficiaries, risk bearers, and parties interested in the project (World Bank 1993). The "units" mainly refer to social organizations, especially those associated with environmental protection. Because those experts somewhat strengthen the professionalism of the environmental protection organizations, they can promote the due decisions in a more effective way when taking part in EIA (Zhu 2009). "Experts" refer to those who have gained knowledge from particular disciplines. With the special knowledge and skills of the experts, the process of EIA can be more scientific. What is more, the experts have no vested interest in the objectives of the EIA, so they can make a fair and objective judgment that will, in the end, promote the quality of the EIA (Bai 2011: 178). "The public" mainly refers to the citizen or dwellers who are affected by the projects or planning, especially the dwellers at the site of construction projects who are directly influenced.

There are no obvious differences concerning the definition of the scope of the public between China's legislation on EIA and universal practice. The key point is that due to the size of the public, not everyone who is willing to participate in the EIA is actually able to participate. Therefore, there is the problem of how to choose the public under different approaches to participation, such as questionnaires, symposiums, demonstration meetings, hearings, etc. For this purpose, the Interim Measures required that "The project owner, the entrusted EIA agencies, or competent departments of environmental protection administration should comprehensively consider factors such as region, profession, professional knowledge background, expression capability, the degree of external influences, etc. and then reasonably choose from those citizens, legal persons, or other organizations whose opinions should be solicited"(Article 15). This means that for the affected public, the only thing they can do is passively waiting and hoping they are lucky enough to be the "lucky" public whose opinions can be solicited. The more crucial point is that current laws and regulations fail to provide any remedies for a person who has not been selected but who has expressed the desire to participate and then questions the procedure or results of the public selection. In practice, the members of the public who question the public participation in EIA are always those who have not been selected, and the reason for this is that the deprivation of their right to be heard does not meet the requirements of due process.

8.3.3 Information Disclosure

Sufficient information disclosure is the precondition for realizing public participation effectively; otherwise, public participation may evolve into a purely procedural step of going through the motions. The Interim Measures fills in the blanks in the LEIA, providing the following detailed stipulations on information disclosure of public participation in EIA. First, the project owner should inform the public of certain information, such as the name and contact information of the EIA agency who is undertaking the assessment work, the working procedures, the main tasks and responsibilities of the EIA, the main issues for public opinions, etc., within 7 days after determining the EIA agency (Article 8). Second, the project owner should announce certain information to the public before submitting the EIR for approval, including the summary of the impact that the construction project may cause on the environment, the main points of the strategies and measures on preventing or mitigating adverse environmental impacts, the key points of the EIA conclusions suggested by the EIR, the approach by which and duration during which the public may consult a simplified edition of the EIR, etc. (Article 9). Third, the competent department of the environmental protection administration should announce relevant information upon the acceptance of an EIR on their government website or through other means convenient for the public, after accepting the EIR for construction projects (Article 13). Fourth, the duration of the aforementioned information disclosure must not be less than 10 days, and it should be guaranteed that the relevant information that has been disclosed is available during the whole period of the soliciting of public opinions (Article 13).

Compared with other parts of public participation, the stipulations on information disclosure in China's legislation on public participation in EIA are the most detailed. However, there are two critical defects. First of all, the degree of information disclosure is not high. The greatest amount of EIA information is stated in the EIR, but China's legislation on EIA only requires making the simplified edition of the EIR public. When the EIR, which has dozens or even hundreds of pages, is concentrated to a few pages, the information provided must also be reduced, so it is unable to meet the need of public participation in EIA (Shi et al. 2011). Next, the exception clause of information disclosure is too broad. According to the stipulations, the information associated with EIAs for projects or planning that "must be kept confidential by the State" will not be disclosed (Article 5), and this includes a wide range of situations. Besides state secrets, this also includes commercial secrets and personal privacy. The administrative organ monopolizes its power of deciding whether or not a piece of information constitutes a state secret, and the judgment standard for commercial secrets is highly flexible. Even worse, if there is controversy over whether or not a piece of information should be kept secret, there is a lack of necessary and open methods for dispute resolution.

8.3.4 Participation Approaches

According to the stipulations in China's current legislation on public participation in EIA, the approaches for public participation include soliciting public opinion, consulting expert advice, symposiums, demonstration meetings, hearings, etc. The Interim Measures regulated different and specific requirements for each approach. For example, the scope of issuing questionnaires should be consistent with the scope of influence of the construction project (Article 19); the minutes of symposiums or the conclusions of demonstration meetings should be worked out and placed on file for future reference (Article 23); and the hearings must be held in public (Article 28).

Regarding the issue of approaches to participation, the biggest problem in China's legislation on public participation in EIA is that the application conditions of each approach and the relationships among them are not clear. In addition, the project owner has a lot of discretionary power over the choice of the approach, and the owner can arbitrarily choose the most favorable approach to organizing public participation. As a result, in practice, the project owner rarely chooses to hold a hearing, which is the most complicated option for soliciting public opinions, while more convenient approaches such as questionnaires and symposiums are thus often adopted (Bai 2011).

8.3.5 Effect of Participation

The final aim of public participation in EIA is letting policymakers make the final decision on the basis of considering all factors through the expression of public opinion. Therefore, the effect of public participation embodies a centralized reflection on the binding force of public opinion to policymakers. According to China's current legislation on public participation in EIA, the binding force of public opinion can mainly be found in the following aspects. First, the project owner should give feedback about the process to the members of the public who bring forward opinions in a proper way. Second, the project owner should consider public opinions seriously and attach the final decision of whether or not the public opinion is adopted to the EIR (Article 17). Thus, it can be seen that the effect of public opinion is just a kind of formal binding force in China. As long as the opinions of the public have been carefully considered, it is not necessary to adopt them. This institutional design is reasonable, which is in line with the basic rule of comprehensive decision-making and democratic decision-making. However, due to the lack of necessary mechanisms for supervision and dispute settlement, the authenticity and rationality of "the final decision of whether or not the public opinion will be adopted" cannot be guaranteed.

According to current provisions, there are two supporting measures for the public opinions being carefully considered. First, the competent department of the environmental protection administration can organize an expert advisory committee to examine the decision related to the situation of adopting the public opinion in the EIR, judge its rationality, and put forward handling suggestions (Article 17). Second, the public can send a report with clear and specific comments in writing to the competent department of the environmental protection administration that is responsible for approval or reexamination, under the circumstances that the public thinks the project owner or the entrusted EIA agencies have not adopted the public opinion, a specific decision is not attached to the report, or the reasons why the public opinion has not been adopted are not reasonable. The competent department of the environmental protection administration that is responsible for approval or reexamination can verify public opinion when necessary (Article 18).

Obviously, such supporting measures are not strong enough. Examination by the expert advisory committee is an internal administrative procedure without openness, while the administrative organ has the power to start the reexamination process at its discretion, and the public's application does not necessarily mean the start of the reexamination process. More importantly, as for the controversy over public participation in EIA, the public does not have access to a more neutral and fair judicial review than the administrative remedies (Wang 2004). The basic reason for this lies in the fact that the rights of the members of the public to express their opinions and to be heard have not been realized or considered as a sort of right by the project owner or administrative organs; thus, naturally, the corresponding appropriate relief has not been provided.

8.4 Latest Progress in Legislation on Public Participation in EIA

8.4.1 Provisions on Public Participation in EIA in the Newly Revised Environmental Protection Law

The Environmental Protection Law was revised in China in 2014 (hereinafter referred to as new EPL), and a new Chapter 5 named "Environmental Information Disclosure and Public Participation" was added. It was the first time that public participation was set as an independent chapter in environmental protection legislation, which reflected that the legislators paid enough attention to public participation. Article 56 in new EPL stipulates the provisions on public participation: "The project owner of a construction project for which an EIR should be prepared pursuant to the law shall explain relevant situations to the potentially-affected public when preparing the report, and solicit public opinions. The competent department responsible for the examination and approval of the report shall public the full text of EIR upon receipt thereof with exception of State secrets or commercial secrets. In the case of a construction project failing to solicit public comments sufficiently, the competent department shall order the project owner to fulfill the task."

Compared with the former legislation on public participation in EIA, there are three major improvements in Article 56 in the new EPL. First, the new EPL expands the scope of information disclosure in public participation in EIA. The former provisions only required the disclosure of the simplified edition of an EIR, but the new EPL requires the disclosure of the full text of an EIR. Second, the new EPL clarifies that the public participation in EIA is the responsibility of the project owner. The former provisions were not clear; as a result, in practice, the department of environmental protection often organized public participation. According to the provisions in the new EPL, the legal obligation for public participation in EIA is associated with the project owner, who can organize public participation by itself or by entrusting it to other agencies; the department of environmental protection has the duty to supervise the public participation in EIA to ensure its authenticity and validity. Third, the new EPL puts forward the demand of the degree of soliciting public opinion as "fully" soliciting public opinion and creates a corresponding supervisory mechanism, that is, if the project owner does not fully solicit the public opinion, the department of environmental protection will instruct the project owner to solicit the public opinion.

8.4.2 Related Provisions in the Measures for Public Participation in Environmental Protection

In July 2015, the Ministry of Environmental Protection (hereinafter referred to as MEP) enacted a supporting rule matched with the Chapter 5 of the new EPL, that is, the Measures for Public Participation in Environmental Protection (hereinafter referred to as Measures) (环境保护公众参与办法), which refines the process and requirements of public participation organized by the department of environmental protection. Compared with the former provisions, the development of the legislation on public participation in environmental protection is reflected in the Measures in the following three ways. First, it makes further improvement in the specific rules for every form of public participation. For example, "When the administrative departments of environmental protection are to solicit opinions through questionnaire surveys, they shall provide an explanation of the basic information on relevant matters. As to the number and scope of people answering the questionnaires, the administrative departments of environmental protection shall take into comprehensive consideration the coverage and extent of environmental influence of relevant matters or activities, degree of social concern, human and material resources required for organizing public participation, and other factors." Second, it initially defines the formal binding force of public opinions, that is, "The administrative departments of environmental protection shall classify, sort, analyze, and research the opinions and suggestions, take into full consideration such opinions and suggestions when making environmental decisions, and give feedback to the citizens, legal persons, and other organizations in an appropriate manner." Third, it requires that "The administrative departments of environmental protection may support and guide social organizations in participating in environmental protection activities by giving financial support for projects, providing purchase services, or other means." Thus, the validity of public participation in environmental protection can be raised.

The Measures shall be applied in the public participation of environmental protection in all aspects. It is an ordinary law on public participation in environmental protection; thus, it can naturally be applied in public participation in EIA.

8.4.3 Progress in the Revision of the Interim Measures

The Interim Measures issued by the former SEPA in 2006 was the most important normative document in the area of public participation in EIA. However, its problems have been fully exposed during its 10-year implementation. In 2014, the MEP commenced the organization of the revision of the Interim Measures, and the revision draft is still in drafting now. In view of existing problems in practice and the new provisions in the new EPL and the Measures, there are three urgent problems to be solved in revising the Interim Measures:

First, the scope of application of public participation in EIA should be defined. In the current administration practice, the legal requirements for public participation are only stipulated in the examination and approval of EIA rather than the whole project, which therefore results in land acquisition compensation, relocation compensation, and other issues related to the projects being put forward in concentration on the public participation in EIA, far beyond the EIA's functional bounds (Zeng 2014). In this regard, the revised Interim Measures should clearly define the scope of public participation in EIA, which means only public opinion about the environmental impact of the EIA object itself, as well as the direct effects on the rights of others due to this environmental impact should be solicited. If the opinions raised by the public have nothing to do with the environmental impact, they do not belong within the scope of public participation in EIA (Wen and Hu 2003).

Second, a clear criterion for "fully" soliciting public opinions should be defined. The vast majority of current disputes being generated in the process of public participation in EIA are due to that the opinions of a large number of the members of the public are never solicited. For that reason, the new EPL requires "fully" soliciting public opinion. The revised Interim Measures should stipulate the criteria for "fully," so that the provisions in the EPL can be implemented. Specifically, whether or not public opinion has been "fully" solicited should be judged from the following aspects: whether or not opinions are solicited from all the people who are directly affected; whether or not opinions are solicited from public representatives generated through an open and fair way when numerous people are directly affected; whether or not the information disclosure in the EIA is comprehensive, accurate, and in due time; whether or not the public has a convenient way to submit opinions; whether or not the period allowed for submissions is long enough; and whether or not public opinions are solicited again through symposiums, demonstration meetings, or hearings for matters with significant differences of opinion among the public.

Third, the effect of public opinion should be defined. Two extreme phenomena are witnessed with the current practice. One is an excessive emphasis on public opinion, that is, any objections or any certain extent of proportion of public opinion will lead to the rejection of the EIR. The other is just soliciting public opinion in order to satisfy the form requirements without considering the real intent of the public opinion. As for these two phenomena, the revised Interim Measures should clearly define that public participation is not a democratic vote, and whether or not the public agrees with one project is not a necessary condition for approving the EIR. The effect of public opinion is mainly reflected in the formal binding force, that is, whether or not public opinion is carefully considered. This is mainly guaranteed through a feedback mechanism, that is, explaining the situation when public opinions are eventually adopted and the reasons when not. And there should also be further clarification that in the event that there is a controversy over the feedback, the public can seek for formal administrative or judicial remedies.

8.5 Conclusion

China's EIA is the product of knowledge gained from the advanced achievements of European and American countries and also has been localized based on China's specific political and economic background. The outstanding performance of this localization is abandoning the process of public participation in EIA. Along with the gradual establishment of a market economy system and enlightened and demonstrated by loan projects from international financial institutions, China began to introduce public participation into EIAs.

From the point of view of legislation, legislation concerning public participation in EIA did not emerge until nearly 10 years after EIA was implemented. However, China's early legislation on public participation in EIA was stipulated too vaguely to be operable. Under the direction of those vague provisions in general guidelines, China's public participation in EIA experienced nearly 10 years of disorder development. Inspired by the Yuanmingyuan anti-seepage engineering incident, China enacted and issued the Interim Measures, unifying and clarifying the basic rules of the public participation in EIA, so that public participation in EIA has achieved an initial standardization.

As the deepening of the practice of public participation in EIA continued, problems like the lack of validity and authenticity began to appear, which presented a new challenge to the legislation on public participation in EIA. Taking the revision of the EPL as an opportunity, China's legislation on public participation in EIA is in the process of revision, and the revision of the Interim Measures has been initiated. Thus, important issues, such as the scope of application of public participation in EIA, the criteria for "fully" soliciting public opinions, and the effect of public opinions, will be regulated and improved.

References

Bai GX (2011) Analysis of the justification of EIA: Take the institution of public participation as an example (环境影响评价的正当性解析——以公众参与机制为例). Journal of Political Science and Law (3): 50

Focusing on Yuanmingyuan anti-seepage engineering (聚焦圆明园防渗工程) People's Daily Online. http://env.people.com.cn/GB/8220/45856/. Accessed 1 Jan 2015

Jin JM (2012) Retrospect of public participation in EIA in past ten years and its future. (公众参与环境影响评价十年回眸与展望) Environmental Protection (22): 37

Li YF (2002) Public participation in EIA (环境影响评价制度中的公众参与). Journal of China University of Geosciences (social sciences edition) (1): 72

Li XM, Li TW (1998) Comparison of public participation in EIA in China and western countries (中西方国家环境影响评价公众参与的对比). Environmental Science (S1): 57

Shi L, Lv HJ, Liu DD (2011) Problems and countermeasures of public participation in EIA in China (中国环境影响评价中公众参与存在的问题与对策). China Population, Resources and Environment (3): 69

Wang J (2004) Research on public participation in EIA: Concurrent discussion on the implementation of relevant provisions in LEIA in China (环境影响评价程序之公众参与问题研究——兼论我国《环境影响评价法》相关规定的施行). Law Review (2): 117–118

Wang J (2006a) A comparative research on EIA institution in China and foreign countries (中外环境影响评价制度比较研究). Peking University Press, Beijing

Wang J (2006b) Environmental law (环境法学). Peking University Press, Beijing

Wen TA, Hu CD (2003) Reflections on public participation in EIA (关于公众参与环境影响评价的若干思考). Present Day Law Science (1): 96

World Bank (1993) Public involvement in environmental assessment: Requirements, opportunities and issues. Environmental Assessment Sourcebook Update, October

Wu J, Cheng SY, Zhang YX, and Shi MM (2012) Searching for the right way from practice: Present situation and development of public participation in EIA (践中取道——公众参与环境影响评价的现状与发展). Environmental Protection (22): 45

Yang L (2013) The EIA for one project in Qinhuangdao was inconsistent with the facts (秦皇岛一项目环评失实). People's Daily, January 29

Zeng J (2014) Conflict and coordination of public participation in EIA for NIMBY facilities: Cite the construction project of passenger dedicated line of Beijing-Shenyang railway as an instance

(论我国邻避设施环境影响评价公众参与的冲突与协调——以北京至沈阳铁路客运专线建设项目为例). Environment and Sustainable Development (5): 125
Zhu Q (2009) Shortcomings and the overcoming methods of individual participation in public decision for environmental protection (环境保护公共决策中个体参与之缺陷及其克服). Law Science (2): 55
Zhu Q (2015) Thoughts and advices on the perfection of public participation in EIA institution in China (我国环境影响评价公众参与制度完善的思考与建议), Environmental Protection (10): 27–28

Chapter 9
Multi-tiered Nature of Environmental Pollution Problems and the Pollution Control Governance in China: The Role of Environmental NGOs

Akihiro Chiashi

Abstract Environmental pollution in China is a multi-tiered problem influenced by many interrelated factors at different scales, such as local, national, and global politics and the economy. Recent ever-increasing environmental pollution incidents in China indicate the limitations of one-sided environmental measures by the government and companies. The role of environmental NGOs and their participation are increasingly significant for improving environmental pollution controlling governance in China.

This chapter studies the multi-tiered nature of environmental pollution in China through case studies in air pollution and industrial pollution. Then, it studies some recent work by environmental NGOs to try to achieve a breakthrough to cope with the politically and economically multi-tiered nature of environmental pollution.

Work by environmental NGOs will have a positive effect on alleviating public discontent concerning environmental pollution. They will also help the government work as an intermediary between the public and polluting companies. Restrictions on this work by environmental NGOs, on the contrary, will worsen the public discontent and prevent progress in the area of pollution controlling governance in China.

Keywords Multi-tiered nature of environmental pollution problems • Pollution controlling governance • Geoeconomic pollution gap • Environmental NGO • Information disclosure • Dialogue

A. Chiashi (✉)
Kyoto University Asian Studies Unit, Kyoto, Japan
e-mail: aki.hiro.0806@gmail.com

H. Kitagawa (ed.), *Environmental Policy and Governance in China*,
DOI 10.1007/978-4-431-56490-4_9

Abbreviations

SEPA	State Environmental Protection Administration
GONGO	Government-operated NGO
NPO	Nonprofit organizations
GC	Green Peace China
Am	Airman
IPE	Institute of Public and Environmental Affairs
GCA	Green Choice Alliance

9.1 Introduction

People in China now have to confront various types of serious environmental pollution in both urban and rural areas. People who live in urban areas, especially in North China, are frequently confronting serious air pollution, while those who live in rural areas are confronting various kinds of industrial pollution, including heavy metal pollution, which causes severe and irreversible health damage to the local people. It is already a well-known fact for the public that some environmental laws, policies, and measures against pollution by the local and central government are still inadequate and untrustworthy in China. The situation is difficult to accurately grasp, and information about the pollution is still unclear for the local people because most of the Chinese environmental policies and governance have been implemented by government agencies that have a strong connection with the polluting companies. The environmental pollution problem in China absolutely needs to develop a new wide and diverse range of governances including third parties, such as NGOs, social media, and citizens' groups.

In spite of the rigid political restriction on activity, some NGOs' remarkable movement against this serious pollution has already begun in some areas of China. Some grassroots NGOs and mass media are trying to find the actual situation of the environmental pollution in their own way and disclose the information to the public to enhance environmental pollution controlling governance.

Since the State Environmental Protection Administration (SEPA) of China enforced "Measures for the Disclosure of Environmental Information (for Trial Implementation)"「环境信息公开办法(试行)」on May 1, 2008, the public has been able to require government environmental bureaus to disclose information about the environment and pollution. Environmental bureaus have to disclose information about environmental laws, regulations, and standards, environmental situations, environmental statistics, a list of violators of pollution standards, etc. Some NGOs are requiring more information about pollution from the polluting companies and local governments utilizing these measures.

There are already many studies on the environmental NGOs in China. However, the research, which has focused on the following questions, is still inadequate. Why is the NGOs' activity necessary and indispensable for the environmental pollution in China? How do each of the NGOs' activities contribute to pollution controlling governance in multi-tiered environmental pollution? Is there any obstacle to NGOs for pollution controlling governance?

This chapter firstly studies multi-tiered environmental pollution in China. Secondly, it studies the NGOs' new movement for "pollution controlling governance" and its role and future difficulties in China.

9.2 Ever-Increasing Environmental Pollution and Its Features in China

People in China have been facing the serious environmental pollution. In recent years, industrial pollution incidents have happened and been made public in various places in China. Heavy metal pollution, which causes serious health damage to the local people, also happens frequently. According to the Ministry of Environmental Protection of the People's Republic of China, there were 11 incidents of heavy metal pollution just from January to August in 2011, and 9 of these were cases of serious lead contamination (Chinanews(中国新闻网) 2011). The lead contamination incidents are called "blood lead" incidents, which describes the exceeding rate of blood lead level of local people caused by the industrial pollution. Many of the blood lead victims are children. Lv and Huang (2014:93–98) indicated that there were 44 incidents of lead contamination from 2004 to 2013. According to the Ministry of Environmental Protection of the People's Republic of China, 4035 people exceeded the blood lead level, and 182 people exceeded the cadmium level as a result of heavy metal pollution incidents that happened in 2009 (China Youth Daily (中国青年报) 2012).

Liu (2010:9) indicated that the number of cancer villages with an unnaturally high rate of cancer is 459 in 29 provinces and direct-controlled municipalities in China. These cancer villages are located along the rivers; industrial water pollution is suspected as the critical reason for the high rate of cancer. Yang (2014:79–91) studied the correlation between water pollution and deaths caused by digestive tract tumors in the Huaihe river basin in an epidemiological study for 30 years. It indicated that severely polluted areas are highly consistent with areas with a high incidence of several emerging digestive tract tumors (Yang 2014:334).

These studies do not reflect the entirety of the industrial pollution situation and the victims' situation in China. Because overall research and epidemiological or medical studies on the victims of pollution have not yet been done in China, the number of past, present, and potential pollution victims is still unclear; however, it is undoubtedly huge.

Air pollution is also a daily threat for the people in China. Year 2013 is now known as the worst year of Wumai (雾霾) in the last 52 years. Wumai is a specific Chinese word to describe the heavy fog and mist that includes various air pollutants. Wumai spread to 25 provinces and more than 100 cities (large and medium scale) in

2013. The average number of Wumai days in China in 2013 was about 30, which is the worst number in 52 years (Xinhua net 2013).

The geoeconomic gap in environmental pollution is also an inevitable issue in China. Zeng (2013) studied the geoeconomic gap in environmental pollution in China. He emphasized that the gap is seen between urban and rural areas, between eastern and western areas, and between rich and poor people in China. The latter are all the more vulnerable to environmental pollution; however, the pollution situation is much worse for them (Zeng 2013:127–159). Villages severely polluted by industrial pollution are mostly in poor areas. This is because heavy or chemical industries tend to transfer from developed to undeveloped or developing areas, such as the inland areas of China.

Zhu (2013:1–4) emphasized that the sources of industrial pollution are transferring from East to Midwest and referred to it as the "pollution relay." Chiashi (2013, 2014) indicated that the sources of chromium pollution in the Eastern areas are mostly already closed; however, they are still operating in the interior regions of China. Some of the sources have caused serious chromium pollution in inland rural areas. Developing areas and people who live there in China are confronting more complicated and serious environmental pollution.

Another critical issue is that the pollution in those areas is structured by politics and the economy at the local, national, and global level. The local people's rights against pollution are strictly restricted and oppressed by the structure.

9.3 Multi-tiered Nature of Environmental Pollution in China

9.3.1 Multi-tiered and Isolated Industrial Pollution

The reason why China needs better environmental pollution controlling governance is the rigidity of the political and economic structure behind the environmental pollution. As many previous researchers have already shown[1], those structures prevent the strict implementation of the environmental laws, policies, and regulations. In particular, the strong collusion between the polluting companies and the local and central government is the crucial obstacle to public participation and multiple governances to environmental pollution by citizens and NGOs.

However, the global structure also needs to be taken into consideration, under the present situation where China is "the factory of the world" in the world economy, in addition to the domestic and local structure behind the pollution. Chiashi (2013, 2014, 2015) found that the global economy is also influencing industrial pollution, in addition to these local and national political economic

[1]See Kitagawa (2008) and Kitagawa (2012). Ren and Shou (2013) also focus on the environmental governance and the problems in China from various perspectives.

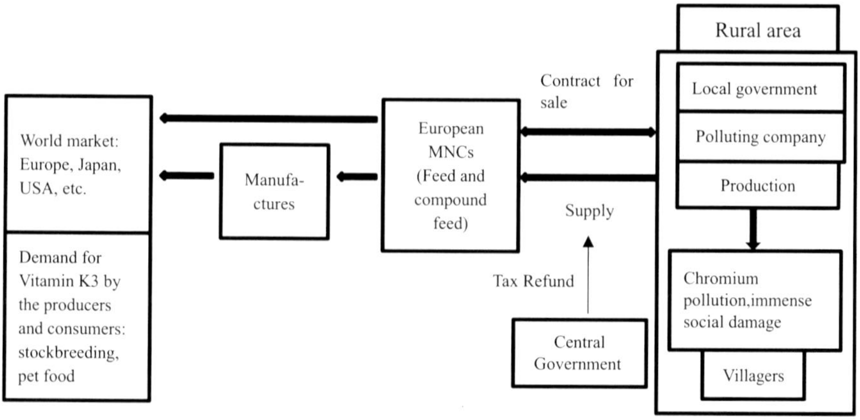

Fig. 9.1 Multi-tiered nature of industrial pollution in China (a case of chromium pollution) (Source: Chiashi 2015, p. 65)

structures, through a case study of chromium pollution in a village in Yunnan province in China (Fig. 9.1).

According to the case study, the polluting company whose main products are chromium salt and vitamin K3 has a strong relationship with the local government through enormous tax revenue. The central government also supports the polluting company's business by tax refunds to promote sales for the global market. Global demand on the products from producers (multinational corporations and manufactures) and domestic and global consumers has also expanded the business of the polluting company. That global demand supports the business of the polluting company, even though the company has performed inadequate treatments for the chromium pollution for several years. Global consumers purchase the polluting company's main product, vitamin K3, which is included in pet (cat and dog) food as an ingredient.

Pollution damage to the local area and people there has spread with the expanding domestic and global business; however, the local people have been hit extremely hard by the pollution damage. In this case, the chromium pollution has remained for more than 20 years, even though the local and central government were made aware of the problem. The local people have had almost no recourse against the multi-tiered pollution.[2]

This domestic and global economic and trade structure has negated the visits and letters from local people for environmental protection facilities, requesting higher cost for corporations to install and operate in the area. The polluting company needs to supply "cheap products" for its expanding domestic and global market without environmental protection. Thus, many cases of industrial pollution in which China

[2]There are some ways to claim the compensation to the pollution source, such as visits and letters or litigation. However, these ways are strongly restricted by politics in China. See Chap. 6 or Stern (2013).

plays the part of "the factory of the world" are structured by local, national, and global systems; however, they are seen as a "local issue." Local pollution victims are isolated and ignored by these global and domestic political economic structures.

9.3.2 *Multi-tiered Nature of Air Pollution*

Air pollution in China is also a diversely structured and complex issue, not only with various kinds of pollution sources but also with trans-regional political economy issues, poverty issues, economic development, and political dynamics behind the pollution.[3] As Fig. 9.2 shows, because China has been the "factory of the world," heavy industries, which are one of the biggest sources of air pollution, such as the iron and steel industry, have intensified with expanding domestic and global demand. In addition to the local and central governments and polluting companies, outsiders, such as multinational corporations (MNCs), and producers and consumers of other countries (especially developed countries) have been influencing the air pollution in China through global business activities.

Another key issue is the air pollution gap between regions in China. For instance, Hebei province has been facing the worst air pollution situation in recent years (Ministry of Environmental Protection of the PRC 2014). Of the 10 cities in China with the worst air quality, 7 in 2014 were in Hebei province. A huge number of

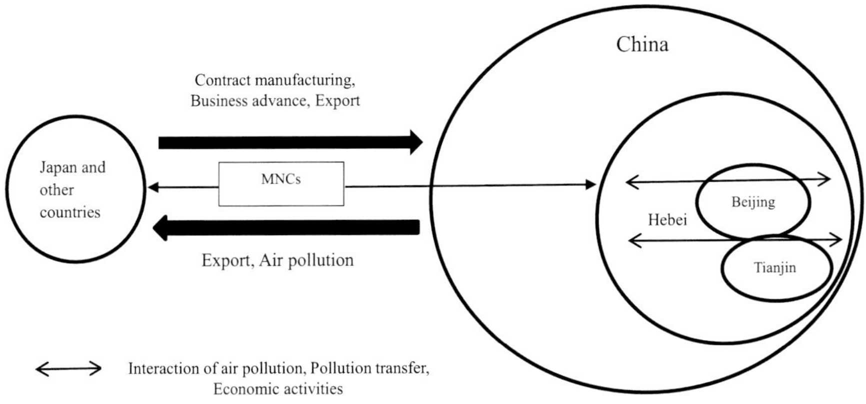

Fig. 9.2 Multi-tiered factor of air pollution in China (Source: Chiashi 2015, p.34)

[3]Chiashi (2015), Chapter 1, studied the worst air polluting region (Beijing, Tianjin, Hebei) during 2013 to 2014 in China. Industrial pollution sources, particularly heavy industries, are gathered in Hebei province, and the situation is worst in Hebei, because of the political and economic reason of the region.

heavy industries and thermal power stations are the main reason for Hebei's air pollution.

The "pollution relay phenomenon" is seen in the Hebei region. Hebei has accepted many polluting industries from Beijing and Tianjin to realize economic development as one of the less-developed areas in the region. The Beijing Olympics in 2008 also influenced the transfer of polluting industries from Beijing to Hebei. One of the slogans of the Beijing Olympics was the "Green Olympics," which enforced the cleanup of large sources of air pollution in Beijing. Nearly 1000 enterprises have transferred from Beijing and Tianjin to Hebei since 2002, and those transferred enterprises were resource consumption industries, such as iron, steel, cokes, metallurgy, machinery manufacturing, etc. (Zhang and Qi (2012:369). Hebei also provides some electricity to Beijing via thermal power stations in Hebei. Here we see the "geoeconomic pollution gap" due to the political dynamics behind economic development and international events (the Olympic games). Global demand for the products also accelerates that gap and dynamics.

These examples show that environmental pollution in China is structured by multiple factors in domestic and global politics and the economy. If the environmental issue in China is regarded just as a domestic or local political issue, these multiple dynamics and the related stakeholders are excluded from bearing any responsibility. This stiffens the incidents of pollution and the isolation of the pollution victims. Improving local, national, and trans-regional governance for the pollution sources by multiple actors is indispensable. Citizens and NGOs are the key actors for this.

9.3.3 Pollution Controlling Governance in China: An Indispensable Institutional Need

Considering the abovementioned multi-tiered environmental pollution situation and the geoeconomic pollution gap in China, the development of pollution controlling governance is indispensable to improve the situation. The word "pollution controlling governance" is the more proper term for describing the situation that Chinese people are seriously facing.

The term "environmental governance" may not be suitable for the current environmental pollution situation in China. The term "environmental governance" in developed countries is frequently used as the governance for global environmental issues like climate change issues in local, national, and global scale.[4] There are many definitions and related arguments about environmental governance. Here it is simply defined as the capabilities, management, and system for conserving the

[4]Recently, many of research papers or books whose title includes "environmental governance" don't focus on industrial pollution. Those researches are mostly focused on the issue of climate change, renewable energy, forest management, sustainable community development, etc.

environment by multiple actors. Environmental pollution has tended to be alienated from the study of environmental governance, because environmental pollution, such as air and water pollution, tends to be thought of as a domestic or local political issue. In other words, the related laws or ordinances and the appropriate implementation by the local and central government can solve the environmental pollution. However, as Japanese experience has already shown, industrial pollution in Japan did not improve at all without participation and governance by multiple actors, such as pollution victims, citizens' groups, lawyers, doctors, independent academics, the media, etc.(Miyamoto 2014).

These multiple actors have been confronting industrial pollution over a long period of time to change the politics, related environmental laws, and polluting companies. This multiple actors' challenge against the pollution, capabilities, and system to tackle the pollution is what I here call "pollution controlling governance," which is the essential factor to improve the environmental pollution in China.

China absolutely needs better pollution controlling governance, not only to change the serious environmental pollution situation but also for the relief of the pollution victims. Diverse support for the pollution victims is essential for their relief, such as medical checks and support, lawsuits, and compensation.

9.4 Environmental NGOs and Pollution Controlling Governance in China

9.4.1 Environmental NGOs in China

Environmental NGOs' participation will be a breakthrough to the multi-tiered environmental pollution in China. There were already 7881 environmental NGOs in China by the end of 2012, which was an annual increase of 38.8 % since 2007; 6816 among these are social organizations (全国生态环境类社会团体) and 1065 are NPOs (生态环境类民办非企业单位) (People's Daily 2013).

The All China Environment Federation (2008:3) sorted and calculated the number of the 4 types of environmental NGOs in China by October 2008. First are the environmental NGOs established by the government, such as the All China Environment Federation, the China Environmental Protection Foundation, the China Association of Environmental Protection Industry, etc. (1309). These are kinds of government-operated NGO (GONGO). Second are the students' social organizations of environmental protection, such as social organizations in schools or union thereof (1382). Third are the grassroots NGOs, which are established by citizens, such as Friends of Earth, Global Village, or nonprofit organizations (NPOs) of environmental protection (508). Fourth are the international environmental NGOs with offices in China, such as Green Peace China (90).

Although the number of NGOs in China is on the rise, these organizations still face restrictions by the Chinese government. Generally, NGOs need to be independent from the government to be a third party. However, in China, all types of NGOs, even the grassroots NGOs, need to legally register with the government division of civil affairs and operate under observation.[5] Environmental pollution issues are a kind of "sensitive issue," which are deeply connected with local or national politics and the economy. Therefore, some grassroots NGOs who directly combat the sensitive pollution issue aren't able to register as formal NGOs but have to register as companies under observation by the government. Many of the grassroots NGOs are confronting various dilemmas between maintaining legal standing and actual activities (Tang (2014:296).

Furthermore, an increasing number of environmental NGOs do not directly operate for the improvement of pollution controlling governance. Most of them are working for environmental education or nature and animal conservation activities, not for industrial pollution and for relief for pollution victims, even if they actually want to act for the latter issues. This is because environmental pollution is still recognized as a politically "sensitive issue" by the government.

9.4.2 Institutional Change for Pollution Controlling Governance

The number of grassroots environmental NGOs is still small and their activity is restricted; however, recently there have been some challenges by NGOs to start a remarkable movement for pollution controlling governance in China. Information disclosure is an indispensable way to improve pollution controlling governance in China. There are two reasons for this.

First is the lack of information about pollution incidents. A number of serious industrial pollution incidents have happened in rural areas; however, investigation and research is still inadequate in China. Local governments are reluctant to do the investigation due to their strong connection with polluting company. Therefore, the NGOs' field survey and provision of information to the public is essential to enhancing public awareness and pressuring the government and the polluting company.

The second reason is the inadequate information and data about the pollution situation at the local and national level. The reliability of the government pollution data is suspected due to the big gap between the data and the actual pollution that the local people are facing. Therefore, the NGOs' own survey as a third party and request to disclose the government pollution information is essential to enhancing the quality and reliability of the pollution information and data.

There were some institutional changes that relate to the disclosure of environmental information. The institution of government information disclosure has

[5]There are some social organizations who work as grassroots NGOs which don't register the government division of civil affairs. They register to division of industry and commerce as a company or work without registration (Huang (2014:91).

changed since 2008 in China. "Measures for the Disclosure of Environmental Information (for Trial Implementation)" (环境信息公开办法(试行)) enforced on May 1, 2008 is one of the key measures for the disclosure of pollution information to the public (Ministry of the Environmental Protection of the People's Republic of China, Measures for the Disclosure of Environmental Information (for Trial Implementation)). Article 4 prescribed that environmental bureaus should promptly and accurately disclose government environmental information, confirming the principle of fairness, equity, convenience for the public, and objectivity. Corporations should promptly and accurately disclose environmental information with the principle of spontaneity and obligation.

Article 5 prescribed that the public, corporations, and other organizations are able to apply for and obtain government environmental information. Article 6 prescribed that environmental bureaus should form a solid environmental information disclosure system.

The Environmental Protection Law of the People's Republic of China (中华人民共和国环境保护法), revised in 2014 and enforced since January 1, 2015, also includes content about public participation and NGOs' activities. Article 53 prescribed that the public, corporations, and other organizations are entitled to obtain environmental information and to participate and observe environmental protection, on the basis of the law.

These measures and laws prescribed that the public is entitled to request and obtain government environmental information. Environmental bureaus should disclose the environmental information and form a system. However, whether the public is entitled to request or obtain corporate environmental information is unclear. The word spontaneity means "corporate spontaneity," and the word obligation means enforcement by the government.

Therefore, the public is able to request the disclosure of pollution information by environmental bureaus on the basis of the law and measure. However, whether a polluting company discloses information about pollution depends on the government's decision or the company's spontaneity. Hence, if public doubt the polluting company's data collected by the government is irrelevant, the company does not have the legal obligation to show the real data to public without government enforcement.

9.4.3 Challenge for Heavy Metal Pollution Controlling Governance in Rural Area[6]

As the third sentence already shows, industrial pollution in rural areas in China is strongly structured by multiple stakeholders, such as the local and central

[6]These NGO's works in this chapter are based on their disclosed materials, website, or interview by author. However, the aim and effect of the work doesn't represent NGO's view but its author's own view.

government, polluting companies, MNCs, and consumers and producers around the world. In this multi-tiered pollution by multiple stakeholders, victims of pollution in rural areas tend to be isolated and ignored by other domestic people and the international society. Disclosing information about the pollution and expanding awareness is one of the key factors to change the closed and isolated situation.

The Changsha Shuguang Environmental Charity Development Center (长沙市曙光环保公益发展中心, Shuguang) in Hunan province in Changsha city was established in August 2013. Shuguang's work is investigation and research for public perspectives on the environment and health, such as drinking water, soil, food, air pollution, water pollution, etc. One of Shuguang's remarkable projects is their own field research concerning heavy metal pollution or water pollution, after which they disclose the information to the public through their website and SNSs (Social Networking System, Weibo, WeChat). Shuguang has already completed several investigations on places with heavy metal pollution, which are called cancer villages, near the mines in Hunan province. Their work is not just for investigation and research but also for creating a dialogue with villagers to provide relief.

They held a "Hunan First Public Meeting for the Relief of Environmental Pollution" at a village on June 27, 2015.[7] The villagers, doctors, freewill public participants, government officials, researchers, and NGO members participated at the meeting to discuss the situation and relief for the pollution victims. This diverse stakeholders' participation has had some positive effects on the multi-tiered and closed industrial pollution.

First is the alleviation of the environmental conflict between villagers and the government. This kind of meeting gives the villagers an opportunity to discuss their situation and future relief with various people. Their discontent and distrust for the government might be alleviated through repeated dialogue. This is because NGOs and other supporters listen to their discontent and try to find a way to provide relief or to be a mediator with the government and the polluting company for the villagers.

Second is the effect of information disclosure. Because industrial pollution is an isolated issue, relieving the pollution victims needs more public attention. Disclosing information about this kind of meeting attracts a more diverse range of public attention to support the villagers. That public attention is also providing positive pressure for the government to provide relief to the pollution victims.

There is another NGO at work on industrial pollution in rural area. Green Peace China (绿色和平, GC) is an international NGO in Beijing. GC also has done some investigative work and research on local environmental pollution and disclosed it to public. Their recent work is field research on Asia's biggest lead and zinc mine at Yunnan Lanping.[8] GC's research and reporting is a valuable information for the

[7]The contents of the meeting can be referred on Shuguang's website (http://www.sggreen.org/index.html, accessed in 6 August, 2016) or Weibo.

[8]Green Peace China, Research on the Asia's largest lead and zinc mine at Yunnan Lanping: Various pain, https://www.greenpeace.org.cn/lead-zinc-pollution-report/, accessed in June 18, 2015

public to learn about the local pollution situation. This is because the information about the pollution in rural areas is considerably restricted.

In the case of the chromium pollution in Luliang county in Yunnan province in 2013, GC not only disclosed their investigation report but also created and disclosed a "chromium prevention manual" for the local people and public to avoid or deal with the chromium pollution.

Thus, these NGOs' projects concerning industrial pollution in rural areas are not only their own investigation but also a practical contribution to pollution controlling governance.

9.4.4 Challenge for Trans-regional Air Pollution Controlling Governance

As we saw in the third sentence, air pollution is a trans-regionally structured issue. NGOs need trans-regional activity, such as trans-regional observation and information sharing about the air pollution situation, pollution sources, pollution relays, etc. There is already cooperation between grassroots NGOs for trans-regional air pollution controlling governance.

The China Air Observation "North China Coal Issue Group" (中国空气观察华北"煤问题"团队) was established by 5 NGOs in August 2013.[9] One of their remarkable projects is their joint fieldwork, research, and information disclosure about the air pollution sources in Beijing, Tianjin, Hebei, and Shandong provinces. These trans-regional research and investigation results, which are based on the "public perspective," were disclosed as the "First Report on the North China Coal Issue" in a workshop and website at Nature University.[10] The project is based on a statistical analysis, fieldwork, and public perspective and advocacy about the government's coal pollution control policy and its effectiveness in 12 cities in four provinces and the direct-controlled municipalities at Beijing, Tianjin, Hebei, and Shandong.

The report points out the questionable appropriateness of the government's pollution-related data, the appropriateness of the reduction goals, etc. For instance, the report points out the ambiguous volume of crude steel production and production capacity in Hebei and Shandong provinces. Furthermore, the report insists that

---

[9]5 NGOs are Greenqilu (绿行齐鲁),Tianjin Luling (天津绿领),the Rock Environment and Energy Institute (磐石环境与能源研究所), China Air Pollution Observation Hebei Volunteers Group (中国空气观察河北志愿者小组), and Nature University (北京水源保护基金会自然大学基金(自然大学)).

[10]Advancing Public Management, Support North China Fog Control: First report on "North china coal issue" http://www.nu.ngo.cn/xxtd/1577.html, accessed in June 12, 2015. Nature University (自然大学, NU) is one of the grassroots environmental NGOs in Beijing. NU's work is not just for the nature conservation but for the research, investigation, information gathering of environmental pollution, and disclosure of the information to the public.

the government's reduction goal for "production capacity" might be meaningless due to the big gap between actual crude steel production and production capability. The report points out that Hebei's crude steel production capability is 286 million tons, while the actual crude steel production was 190 million tons, both in 2013. Hebei's crude steel production capability reduction goal is 60 million tons to be reduced by 2017. However, if it is achieved by 2017, the capability will be 226 million tons, still over the 2013 actual production volume (190 million tons). Therefore, they insist that Hebei's crude steel production capability reduction goal must replace the production reduction goal.

This kind of "public perspective report" by the NGOs and volunteers' surveys and research enhance public awareness and knowledge about the government data, policies, and pollution reduction plans.

This North China coal issue project has already ended. However, there is another remarkable trans-regional project called "Airman" (好空气保卫侠, Am) which was started in September 2013 from an environmental activists' fund. This project shares air quality information that is collected by volunteer members of the group in Chinese SNS (Weibo or WeChat) and continuously observes the air quality and pollution sources in various places. When they find serious, violating air pollution sources, they report them to the local government and public via SNS. Am members support pollution victims, even for public interest litigation. Am tries to observe the government and polluting companies and urge the government to disclose air quality information (China Philanthropist, 2014). Am members are gradually increasing in number,[11] which means public awareness toward air pollution and pollution sources is gradually expanding through the SNS.

Furthermore, some other NGOs' projects also contribute to air pollution controlling governance in China. Zhen and Yin (2013:175–187) studied environmental NGOs' participation and contribution to the PM2.5 incidents in China. They concluded that the NGOs' activity, such as publicity campaigns, data collecting and monitoring, and disclosures on the Internet, has contributed to the enhancement of public awareness and the improvement of government measures.

Thus, the NGOs' trans-regional activities are gradually enhancing public awareness and participation and are spurring government reactions to air pollution through public perspective investigations and information disclosure.

9.4.5 Challenge for Globalized Environmental Pollution Controlling Governance

The Institute of Public and Environmental Affairs (IPE, 公众环境研究中心), established in May 2006 in Beijing, is one of the most unique and famous NGOs

[11] According to the interview to an Am organizer on November 3, 2014, at Beijing, followers to Am are about 300.

whose activities have accelerated domestic and global pollution controlling governance. One of the IPE's representative projects is information disclosure about the environmental pollution sources in China. The IPE collects the available pollution data disclosed by the government in every location in China and discloses it via a database on their website as a "Pollution Map."[12] This map not only encourages polluting companies to carry out pollution control but also gives the pollution information to the public in various places. Foreign companies' pollution data is also on this pollution map.

Because of the legal restriction on the public's ability to obtain pollution source data, the IPE's pollution data is all collected by government data sources. However, the map is realized easy access to the pollution data, enhancing public concern and putting pressure on the polluting sources and local governments. Even if a company who has violated environmental standards reacts and objects to the IPE's data, the IPE has a channel for dialogue with the pollution source. This activity is a kind of breakthrough for the public to the closed pollution information within companies and the government.

Another of the IPE's activities is the "Green Choice Alliance (GCA)" for responsible supply chain management programs on a global scale. The GCA program involves activities by an alliance of NGOs, which tries to build "global supply chain governance" utilizing the IPE's pollution database and disclose the information (IPE 2008). Because China has been the factory of the world, many subcontracted factories (upstream in the supply chain) have intensified in China as pollution sources. Domestic and foreign makers (downstream in supply chain) buy their products or components from subcontracted factories and sell them to the domestic and global market. However, as seen in the third section, the responsibility of buyers and consumers amounts to nothing at all in the global economy system. The GCA program tries to improve the system through information disclosure about pollution at subcontracted factories in China. The GCA encourages makers and consumers to make the green choice, which means the boycott of non-supply chain management for downstream buyers' products. It evaluates and rates the downstream buyers (mostly world-famous brands) by GCA's own criteria, such as communication and follow-up, compliance and corrective action, green supply chain practices, data disclosure, responsible recycling, etc.

One of the GCA's remarkable achievements is the case of world-famous IT Company Apple. The IPE and other NGOs investigated and published the situation of Apple's subcontracted factories' pollution in China in April 2010. Since then, the IPE and Apple have had several constructive dialogues to improve Apple's subcontracted factories' pollution (Friends of Nature, IPE, Envirofriends, Nature University and Nanjing Greenstone 2013). The pollution databases have proven valuable for MNCs looking to monitor their suppliers in China (Tan 2014:56).

[12]Pollution map is available on the IPE website, http://www.ipe.org.cn/pollution/index.aspx, accessed in August 6, 2016. The data is renewed all the time.

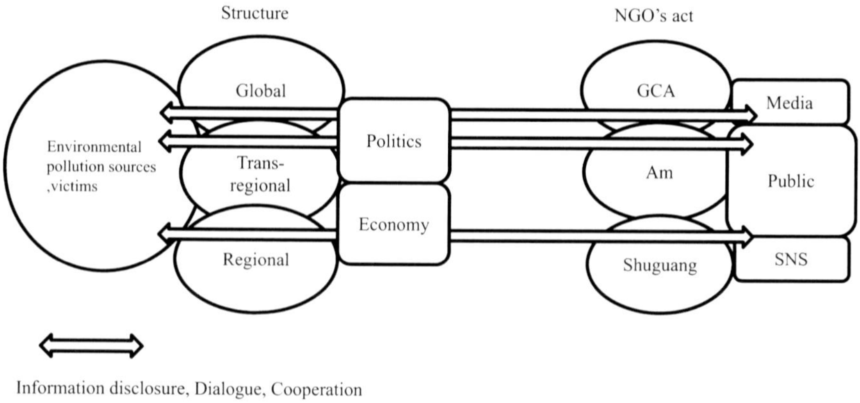

Fig. 9.3 NGO's role for the multi-tiered and isolated environmental pollution

The GCA program has expanded their work for the textile industry, food and beverage industry, household and personal care industry, automobile industry, brewery industry, paper industry, and leather industry (IPE and Natural Resource Defense Council 2014). This GCA work has the potential to boost cooperation between international and domestic companies and NGOs and contribute to the building of the global supply chain pollution information disclosure system.

In other words, the GCA program is a breakthrough into globally structured environmental pollution as an open information system that encourages public participation and pollution controlling governance at the global level. This work now involves more stakeholders around the world and encourages them to take more responsibility for the pollution in China.

Figure 9.3 describes the role of environmental NGOs in pollution controlling governance in China. These three NGOs are all trying to access pollution sources, victims, and government by the way of the public perspective. There are multi-tiered obstacles of politics and economy at the regional, national, and global level. The three NGOs' common way of overcoming the obstacles is information disclosure about pollution. They collect information about pollution sources and victims in their own way and provide this to the public through the media and SNS. They occasionally have dialogues with pollution sources, the government, victims, and the public. That will enhance the government, pollution sources, and public awareness of pollution issues. The number of participating stakeholders is increasing as the NGOs' work develops. The work by these NGOs will also help the government, such as serving the role of intercession between government and pollution victims, lowering the risk of trouble between polluting companies and residents, etc.

9.5 Conclusion

Politically and economically multi-tiered and isolated environmental pollution incidents in China need more public attention to break through to the situation. Some grassroots NGOs are already challenging for pollution controlling governance with new ideas and measures in China, even if their activity is still legally and politically restricted and their budgets are weak. Their work is indispensable for improving domestic geoeconomic pollution gap, relief for the victims of pollution, and local, trans-regional, and global pollution controlling governance.

The local, environmental grassroots NGO of Shuguang is trying to do field research about villages with heavy metal pollution and disclose the information to the public by WeChat, websites, and the media (TV and newspapers) or regularly holding open workshops for the public. These activities are attempts to convey more of reality to the public and to positively pressure the government and companies to take pollution control action. This is actually providing some better results for local pollution controlling governance. Trans-regional work for air pollution and the GCA are also remarkable movements to enhance pollution controlling governance, particularly improving the dialogue between the public, governments, and polluting companies.

However, there are some future problems for the improvement of pollution controlling governance in China. The first problem is the lack of independent doctors' participation in pollution controlling governance. As Japanese experience shows, doctors' medical examinations are necessary to prove the causal relationship between pollution victims and industrial pollution. Some academics, mainly in natural sciences fields, are already participating and assisting the NGOs' research in China. However, doctor participation is still rare in all cases of industrial pollution in China. The main reason is that industrial pollution is still a "sensitive issue" with connections to the local and national politics and the economy.

The second problem is that the attention on grassroots NGOs by the international society is still weak. Awareness toward environmental pollution issues in China from the outside is essential in order to urge polluters and the government toward pollution control. International cooperation with grassroots NGOs will be a key factor to break through the multi-tiered environmental pollution. The IPE is already trying to connect with international society in some ways, like the GCA project. However, even for a famous NGO IPE, the number of cooperative companies and consumers is increasing but still limited, and attention or support for the work of the IPE from the outside is still limited. Shuguang is a local grassroots NGO whose attention and cooperation from outside is less than IPE.

The third problem is the restriction and observation of the NGOs' activities by the government. Grassroots NGOs are now confronting a dilemma between maintaining legal standing and actual activities with political restrictions. Recently, the restrictions have become stronger, in contrast to improvements to the related environmental laws. Those strong restrictions will prevent constructive pollution controlling governance in China and keep the political economy behind the pollution in place. It is clear that the work of NGOs is contributing to pollution controlling governance, such as pollution monitoring, pollution victims' relief, trans-regional information sharing, etc. NGOs can be an intermediary between the public, the government, and polluting companies. These NGOs' works and roles as third parties will have a positive effect on alleviating public discontent concerning environmental pollution. That will also help the government work.

The responsibility of international society is to observe the restrictions carefully and then try to find a way to cooperate with grassroots NGOs confronting the dilemma.

References

All China Environment Federation (2008), 2008 Report of the Environmental NGO in China, All China Environment Federation (in Chinese) 中华环保联合会《中国环保民间组织发展状况报告》中华环保联合会.

Chiashi, Akihiro (2013), Industrial Pollution in China and the Global Economy: A Case Study of Xing Long Village, Yunnan Province, Proceedings of the 5th Next Generation Global Workshop:Social Innovation and Sustainability for the future. Global Center of Excellence for Reconstruction of the Intimate and Public Spheres in 21st Century Asia, Graduate School of Letters, Kyoto University, pp.500–511(in English).

Chiashi, Akihiro (2014), Various Aspects of Chromium Pollution in Contemporary China: Pollution Damages and the Structural Factors. People and Environment, Vol.40, No.1, Japan Association on the Environmental Studies, February 2014, pp.19–32 (in Japanese).

Chiashi, Akihiro (2015), The Political Economy of the Environmental pollution in China, Showado (in Japanese).

China Philanthropist, April 2014, "Airman", http://news.hexun.com/2014-04-16/163981497.html, accessed in 6 August, 2016.

China Youth Daily (中国青年报) 1 Feb, 2012 "Heavy metal incidents frequently happens", http://zqb.cyol.com/html/2012-02/01/nw.D110000zgqnb_20120201_5-07.htm, accessed in 6 August, 2016.

Chinanews(中国新闻网), 25 October, 2011, "Ministry of Environmental Protection: 11 Heavy Metal Pollution Incidents Have Happened This Year from January to August". http://www.chinanews.com/cj/2011/10-25/3413470.shtml,accessed in 6 August, 2016.

Friends of Nature, IPE, Envirofriends, Nature University and Nanjing Greenstone (2013), Apple Opens Up: IT Industry Supply Chain Investigative Report-PhaseVI, http://114.215.104.68:89/Upload/Report-IT-Phase-VI-EN.pdf, accessed in 6 August, 2016.

Huang, Mei (2014), Institutions, Tsujinaka, Y., Li, K., Kojima, K. ed.,Civil Society and interest group in contemporary China:China in a comparison, Bokutakusha, pp. 85–109 (in Japanese) .

IPE (2008), The Green Choice Alliance for responsible supply chain management program, http://114.215.104.68:89/Upload/Report-GCA-Brochure-EN.pdf, accessed in 6 August, 2016.

IPE and Natural Resource Defense Council (2014), CITI Index and the Initial Assessment of 147 Brands: Greening the Global Supply Chain, http://114.215.104.68:89/Upload/IPE-Reports/Report-CITI-EN.pdf, accessed in 6 August, 2016.

Kitagawa, Hideki (ed.) (2008), Environmental Problems and Law, Policy in China: towards sustainable development of East Asia, Horitsu Bunka Sha (in Japanese).

Kitagawa, Hideki (ed.) (2012) Chinese Environmental Law and Policy and Governance: Current status and Problems of Implementation, Koyo shobo (in Japanese).

Liu, Lee (2010) "Made in China: Cancer Villages", Environment, 2010, Vo.52, Taylor& Francis, Philadelphia, pp.8–21.

Lv, Zhongmei and Huang, Kai (2014), It is urgent to make the Environment and Health Law from 63 Environmental Incidents, Friends of Nature, Liu Jianqiang ed., Annual Report on Environment Development of China, Social Science Academics Press (China), pp.92–108 (in Chinese). 吕忠梅· 黄凯(2014)《从63个案例看环境与健康立法迫在眉睫》刘鉴强· 自然之友编《中国环境发展报告(2014)》社会科学文献出版社。

Ministry of Environmental Protection of the PRC: Air quality of key areas and 74 cities in 2014, (2014) http://www.mep.gov.cn/gkml/hbb/qt/201502/t20150202_295333.htm,accessed in 6 August, 2016.

Ministry of the Environmental Protection of the People's Republic of China, Measures for the Disclosure of Environmental Information (for Trial Implementation), (2015) http://www.zhb.gov.cn/info/gw/juling/200704/t20070420_102967.htm, accessed in 29 May, 2015.

Miyamoto, Ken'ichi (2014) A Historical Treatise of Industrial Pollution in Japan after World War (Sengo Nihon Kogaishiron), Iwanami Shoten (in Japanese).

People's Daily, 5 December, 2013 "Around 8000 environmental NGOs in China", http://paper.people.com.cn/rmrbhwb/html/2013-12/05/content_1357312.htm, accessed in 6 August, 2016.

Ren, Bingqiang, Shou, Huisheng ed., (2013), Chinese Environmental Governance: Dynamics, Challenges, and Prospects in a Changing Society, Palgrave Macmillan.

Stern, Rachel (2013), Environmental Litigation in China: A study in Political Ambivalence, Cambridge University Press.

Tan, Yeling (2014),Transparency without Democracy: The Unexpected Effect of China's Environmental Disclosure Policy, Governance: An International Journal of policy, Administration, and Institutions,Vol.27, No.1,Willey Periodicals, INC, pp.37–62.

Tang, Xiumei (2014), International Society and Grassroots NGO, Tsujinaka, Y., Li, K., Kojima, K. ed., Civil Society and interest group in contemporary China:China in a comparison, Bokutakusha, pp.289–303 (in Japanese),.

Xinhua net, 30 December, 2013, "Year 2013, China faced the worst Wumai in 52 years", http://news.xinhuanet.com/city/2013-12/30/c_125930861.htm, accessed in 6 August, 2016.

Yang, Gonghuan (2014), Correlation between Water Pollution and Deaths Caused by Digestive Tract Tumors in Huaihe river Basin, Friends of Nature, Liu Jianqiang ed., Annual Report on Environment Development of China, Social Science Academics Press (China), pp.79–91 (in Chinese). 杨功焕(2014)《淮河流域水环境污染与消化道肿瘤死亡》刘鉴强· 自然之友编《中国环境发展报告(2014)》社会科学文献出版社。

Zeng, Jianping (2013), Environmental Justice in Present-Day China, Social Science Academic Press(China) (in Chinese) 曾建平(2013)《环境公正:中国视角》社会科学文献出版社。

Zhang, Gui, Qi, Xiaoli (2012), Building Hebei Coastal Economic Zone and Pushing Forward Regional Integration,Wen Kui, Zhu Erjun ed.,(2013) Beijing-Tianjin-Hebei Regional Integration Development Report, Social Science Academic Press, pp.350–387(in Chinese). 张贵· 齐晓丽(2012)《建设河北沿海经济带与推进区域一体化》文魁· 祝尔娟编《京津冀区域一体化发展报告(2012)》社会科学文献出版社。

Zhen, Lin, Yin, Guan (2013), Environmental NGOs and Participative Governance: the Case of the PM2.5 Incident, Ren, Bingqiang and Shou, Huisheng ed., (2013) Chinese Environmental Governance: Dynamics, Challenges, and Prospects in a Changing Society, Palgrave Macmillan, pp.175–187.

Zhu, Haizhong (2013), Environmental Pollution and Rural People's Environment Protest: A study Based on Case of N Village in North Jiangsu Province, Social Science Academics Press (China) (in Chinese). 朱海忠(2013)《环境污染与农民环境抗争-基于苏北N村事件的分析》社会科学文献出版社。

Chapter 10
Environmental NGOs and Environmental Pollution in China

Yasushi Aikawa

Abstract In China, environmental NGOs have developed over the 20 years despite its strict political system. This chapter describes the types of environmental NGOs in China and their developing and growing process. It also considers the importance of the lessons learned from the Japanese in association with the power of civil society including NGOs in resolving environmental pollution problems. What are the Chinese environmental NGOs doing in dealing with environmental pollution? How do they make relationships with the Japanese counterparts? The objective of this chapter is to answer these questions. In answering these questions, important lessons can be drawn in mitigating environmental pollution.

Keywords NGOs • Environmental pollution • International exchange and cooperation • Japan • Civil society

Abbreviations

ACEF	All-China Environment Federation
CAN	Climate Action Network
CBO	Community-based organization
CLAPV	The Center for Legal Assistance to Pollution Victims
CSO	Civil society organization
FON	Friends of Nature
GCA	Green Choice Alliance
GONGO	Governmental NGO

This article is a synthesis and update of my past published articles in Japanese, especially Aikawa 2008a, b and 2012. They are mainly based on participant observation of Japanese and Chinese environmental NGOs' exchange projects and interviews with persons related to environmental NGOs in China, which were held in 1994, 1996 and from 1998 to now. Their updating was supported by JSPS KAKENHI Grant Number 25340146.

Y. Aikawa (✉)
Faculty of Business Administration, Tottori University of Environmental Studies, Tottori, Japan
e-mail: aikawa@kankyo-u.ac.jp

H. Kitagawa (ed.), *Environmental Policy and Governance in China*,
DOI 10.1007/978-4-431-56490-4_10

GVB	Global Village of Beijing
IPE	Institute for Public and Environmental Affairs
JEC	Japan Environmental Council
NGO	Nongovernmental organization
NPO	Nonprofit organization
PO	People's organization
WSSD	World Summit on Sustainable Development

10.1 Introduction

Over the last two decades, nongovernmental organizations (NGOs) have taken great changes in the environmental field in China. These organizations are organized to solve environmental problems voluntarily. They have developed simple actions into strategic and efficient ones. These NGOs emerged in the environmental area in China in 2003, growing from an interesting topic of conversation.

It was revealed that severe environmental pollutions had taken places in several places in China consecutively during 2004–2005, and in some cases these issues were revealed by NGOs. Before the 2010s, Chinese environmental NGOs wanted to learn the experiences from the Japanese environmental pollution victims, supporters, and NGOs. However, interestingly, Japanese NGOs had also learned from the Chinese NGOs in some cases from 2008.

This article consists of four parts: the first part (Sect. 10.2) provides an overview of the types of environmental NGO efforts and operations in China. The second part (Sect. 10.3) describes a brief history of the domestic grassroots environmental NGOs in China – the most important type of NGOs evolve over time. The third part (Sect. 10.4) discusses the developing process of their actions in tackling environmental pollution. The fourth part (Sect. 10.5) examines the exchange programs and cooperation between the civil societies of China and Japan in relation to environmental pollution.

10.2 Overview of Environmental NGOs in China

The word NGO in this chapter means an organization that is in name only or by institution independent from the government, and whose activities are basically nonprofit in nature. There are several similar words, i.e., nonprofit organization (NPO), community-based organization (CBO), civil society organization (CSO), and people's organization (PO), which are hereinafter referred to as NGO.

China is a socialist country governed by the communist party. As a result, many foreigners, especially those in the Western countries, believe that there is little space for NGO activities in China. However, as a matter of fact, there are many NGOs prevail in China. These NGOs can be divided into two classes based on their origin: international and domestic. The domestic NGOs can be further divided into two categories based on their origin, that is, whether they are governmental or grassroots.[1]

Governmental NGOs, an abbreviation for governmental-nongovernmental organizations, sounds like a contradiction, but they play a major role in China. In fact, similar organizations can be found in many countries in East Asia, but in China the situation surrounding NGOs is the most extreme. It is said that the number of NGOs whose operations are related to environmental issues is in thousands and most of them are governmental NGOs (commonly referred to as GONGOs). GONGOs have their own historical and political social raison d'être. In this chapter, the relationship between GONGOs and grassroots NGOs in the environmental area will be discussed. It is easy to imagine that these two categories of NGOs are rivals, and indeed, they are. But the situation is more complicated than that. For example, there is a relatively new organization that was established in the late 2000s, the All-China Environment Federation (ACEF), which can be regarded as the most important environmental GONGO. They introduce themselves as a "nonprofitable civil society organization (CSO)," and their core members are listed on their website (www.acef.com.cn), where we can observe the fact that the majorities are members of the government or the communist party. That proves that this NGO is a GONGO. On the other hand, many core members of grassroots NGOs also participate in this organization. Several events like parts of the annual conferences are completely handled by these members. This fact means that GONGOs provide the grassroots NGOs with the opportunities for taking action. On the other hand, because the climate change issue is regarded as an international issue in China, it may not be easy for ordinary people, including members of grassroots NGOs, to offer their opinions on such issues as their views may differ from the official positions of the government. However, while grassroots NGOs have interacted with foreign counterparts, they have often been asked to express opinions in the perspective of climate change. To avoid the risk of government animosity resulting from offering controversial opinions, they established a network of NGOs, the CCAN, whose name stands for Chinese CAN,[2] and let the other important GONGO, China Association for NGO Cooperation (CANGO), whose mission is to develop civil society in China, to participate in CCAN as a core member organization. However, as a matter of fact, CANGO often participates as a leading organization.[3]

[1]Many previous researches seem to premise that there are same divisions among environmental NGOs in China as I wrote in this article, like Economy 2005, Turner and Lv 2007, and Bao 2009.

[2]Climate Action Network; in the climate change area, CAN is the most significant worldwide network of NGOs; the number of member NGOs is over 950. CCAN is the official branch of CAN in China; the number of member NGOs is 18. Website: www.climatenetwork.org.

[3]This paragraph is based on participant observation of Japan and China environmental NGOs' exchange projects since 2010 and interviews with persons who related to environmental NGOs in Beijing and Tianjin in August and September 2014.

To distinguish them from GONGOs, grassroots NGOs were previously called the "real" NGOs,[4] which made the staff and members of GONGOs unhappy. Subsequently, in order to exhibit their voluntary action and independence from the government, they chose the word "grassroots." Indeed, "grassroots" means locals literally, but the core members of grassroots NGOs were those initial intellectuals who ranked very high in the hierarchy of the Chinese society and those who would not be regarded as people at the local level. In 1999, these grassroots NGOs initiated activities to bring relief to pollution victims. In the meantime, they approached and began to voice the needs of the local Chinese society, and they (and their activities) seemed to act much more like grassroots organizations than before. Since then in some cases, local people including pollution victims have also established NGOs. Environmental NGOs are one of the pioneering grassroots NGOs. Until now, the environmental area has occupied the most significant and developed area among the fields in which grassroots NGOs operate.

Foreign based international environmental NGOs, including multinational organizations such as WWF and Greenpeace, are also operating in China. Japanese NGOs also represent one kind of these international environmental NGOs with one or two of them being organized by the Japanese citizens in China. They are different from other developed countries' NGOs, especially Occidental countries' NGOs, as described in Sect. 10.5.2. Some of the Japanese NGOs' which exchange projects and cooperation with the Chinese NGOs have successfully and significantly prevented environmental pollution in China. Among all of the NGOs mentioned above, the most important and interesting NGOs in China are domestic grassroots NGOs, which will be the focus of discussion in the following two sections.

10.3 Brief History of Grassroots Environmental NGOs in China

The history of grassroots environmental NGOs in China began in the 1980s, but their activities were interrupted by a historical incident. However, after the United Nations Conference on Environment and Development (UNCED) held in 1992, they resume their activities.

In the 1980s, even junior high school students had heard about Greenpeace and tried to establish similar organizations, and some older students succeeded in establishing NGOs. But several founders of NGOs had ties to the Tiananmen Square protests in 1989, and after the incident on June 4th, all of them were disbanded forcefully.[5] In the early 1990s, several people tried to establish

[4]Compared to GONGOs, the grassroots NGOs could be defined as "real" because their activities and origins are similar to general NGOs in other countries.

[5]One of the junior high school students at that time revealed these facts in 2007. Considering political risk, it should be too early to write who the one is.

environmental NGOs, and one of the candidates, whose aim was to conserve specific living species, got permission to establish one. This NGO[6] was established before the period when the aim of NGOs was getting permission in tackling environmental problems. The first established NGO performed in solving general environmental problems was Friends of Nature (FON) in Beijing. The preparatory meeting for establishment was held in 1993, and it was established and obtained permission in the following year.

There has been more than 20 years since the establishment of NGOs in China, and this period can be divided into two stages: the first stage is from 1993 to 2002, and the second stage from 2003 to the present day. Also, the first stage can be divided into the first phase (1993–1997) and the second phase (1998–2002), and the second stage can further be divided into the third phase (2003–2007) and the fourth phase (2008–present) based on their characteristics.

In the first stage, the typical characteristics of the developing country's NGOs could also be noticed in China's grassroots environmental NGOs. During this period, they displayed the following three common characteristics: (1) their social influence was small; (2) the scope of the problems they wanted to solve was limited to domestic issues; and (3) their financial resources mainly depended on funds from overseas. During this period, the very existence of these "real" or grassroots NGOs in China was seen as significant despite the strict control of China's government. However, their influence was limited.

In the first half of the decade, the first phase, from 1993 to 1997, general organizations were established in Beijing and Yunnan, whose core members were intellectuals. In the second half of the decade, the second phase, from 1998 to 2002, their actions, active areas, members, types of organizations, specialties, and established locations become more diversified. These changes added reality to some parts of the activities by the nominal grassroots NGOs. Today there are five types of founders and origins of NGOs: type 1 are groups of general intellectuals or journalists; type 2 are groups grown from student groups; type 3 are groups of specialists; type 4 are groups of victims of environmental issues and their supporters; and type 5 are founded by experienced persons who have worked or participated in other environmental NGOs before. Almost all of the NGOs established in the first phase were type 1, and all of the other four types appeared in the second phase, though type 5 was exceptional at that period and increased in the third phase, which has been stated at the second paragraph of Sect. 10.3.3. They also developed networks among themselves in the second phase.

The second stage started from 2003 to the present date (2015). In this stage, the typical characteristics of the NGOs in the developing countries including the above (1)–(3) have been lost or are changing. This stage can be described as the transition stage to a developed country's type of NGOs, although the goal is unreachable in the present situation. In this stage, they maintain their general networks on the one

[6]The name of the NGO is Saunders' Gull Conservation Society of Panjin City (盘锦市黑嘴鸥保护协会). Its website is www.heizuiou.com, but it has no English information except the name of the organization at last of 2015.

hand, and make their new networks separate from the old ones on the other hand as new objects, issues, or goals.

The first 5 years from 2003 to 2007, is the third phase, and 2008 and later is the fourth. In the third phase, the grassroots NGOs exhibited the following characteristics (1) change: their influence became broader in that when they revealed facts about some environmental pollution or problems, even major media reported them, and more people, including the highest levels of leaders, watched, read, or listened to them, in some cases making them to take action. Regarding characteristics (2) and (3), a few exceptions could be seen before 2008. Even now, these exceptions still remain as exceptions or minor cases, but the cases increased rapidly after the Olympic year of 2008, with one of the causes attracting the concern of foreigners. At least there is no doubt that it is no longer true that the existence of "real" or grassroots environmental NGOs in China itself is enough to be significant. Their actions and influence are significant as well currently.

10.3.1 The First Phase (1993–1997): The Emergence of the Big 3 in Beijing

FON and the Global Village of Beijing (GVB) are famous around the world as "real" Chinese environmental NGOs since their establishment. Three organizations, the two NGOs mentioned above along with Green Earth Volunteers, made their presence felt and performed their influence until the middle of the 2000s. In the first phase, several "real" environmental NGOs were also established in Yunnan province, which were influenced by the activities of Occidental and international NGOs.

FON took a long time not only in the preparation for its establishment but also for trial and error before their activities were fulfilled. However, foreigners from Western countries thought highly of it and awarded it with some prizes before it had really accomplished anything for it was the first "real" NGO in the environmental area in China. They are symbolic that the existence of "real" or grassroots environmental NGOs is independent from the government or the Chinese communist party, which were valuable enough in this phase and at this stage.

10.3.2 The Second Phase (1998–2002): The Rise of Grassroots NGOs with Multiple Organizational Structures and Characteristics

From 1998 to 2002, many "real" environmental NGOs with diversified characteristics emerged in many locations throughout China, and all of them kept reforming to reflect the diverse environmental needs embedded in the Chinese society, such as transferring their activity bases to various areas, extending activities to cover various specific fields, reflection of minor ethnic groups' requirements, relief for environmental pollution victims added to the objective of activities, etc. The networking is also developed, and the presence of them as a whole was strengthened. This enabled them to change into progressive grassroots organizations.

The transformation of the NGOs that were independent from the government or the Chinese communist party went through the process of being called "real" NGOs to grassroots NGOs, which was the period of preparing for participating in the World Summit on Sustainable Development (WSSD) held in Johannesburg, South Africa, in 2002. As their number increased and their diversity flourished, they wanted to learn and exchange information with each other, and their efforts naturally established a connection among them. In particular, the leadership and the initiative of the GVB were important with regard to two events that were prepared for, not only for themselves but also for the whole environmental civil society of China. The first event was the Earth day in 2000. Indeed, it originated from the USA, which was a perfect nongovernmental campaign in 1970, and it prevailed globally. In China, however, it had been held as a governmental event before 2000; therefore, GVB wanted to change it into a nongovernmental event on the global standard. The second event was WSSD. GVB asked the ways of taking part in it, along with many other environmental NGOs which were independent from the government or Chinese communist party. And they decided to call themselves grassroots (environmental) NGOs during the period of the preparation for participating in WSSD.

The development of the Internet in China in the recent years also accelerated the improvement of their networking. In fact, at that time, compared with the real NGOs, virtual NGOs on the Internet were easier to organize. Therefore, young people from teenagers to the thirties established several virtual NGOs. And if they succeeded, they would grow up into the real NGOs. The Internet not only enabled them to know and understand each other but also provided an opportunity for them to specify their activities.

It was a good opportunity for them to take part in WSSD for it could confirm their identity and build confidence. They had few members to be recognized by participants from other countries. For example, most of the participants from the Japanese NGOs did not notice that their Chinese counterparts had participated in WSSD. But they were not only participating in the conference but also holding a briefing session after returning from abroad to invite other grassroots environmental NGOs that could not take part in the conference. As a result, their identity and confidence were passed on to almost all grassroots environmental NGOs in China at that time.

10.3.3 The Third Phase (2003–2007): Environmental Information Disclosure and Networking

In the first stage or in the first and the second phase, even the very existence of a "real"/grassroots environmental NGO in China was significant, especially toward people in the vicinity of the place it was established, except places such as Beijing, where there were two or more of those organizations. However, beginning in 2003, many grassroots environmental NGOs proved their raison d'être by revealing environmental problems with detailed information and raising issues emerging in succession.

The other phenomenon that began to occur in 2003 was "spinning-off" or "spinning-out." Project leaders or core members left the original organizations like FON or GVB and established new organizations, and began to lead them as leaders.

Networking also changed partially. Until WSSD, networking itself was regarded as a goal. After several years, GVB continued general networking, and later FON also began general networking and had continued publishing annual reports till now. Importantly, after WSSD, they made good use of their general network to ask for cooperation when one NGO wanted to solve a problem with other NGOs, and if there were NGOs that agreed with their request for cooperation, they made a special network, which can be called an objective network. Successful objective networks had social influence, like requiring an environmental impact assessment for setting the sheet pond of the Old Summer Palace (圆明园) in Beijing and the anti-construction campaign against dams on the Nujiang River in Yunnan. The opinions of these networks were broadcast by the mass media, and environmental pollution problems were revealed by the local NGOs as described below. Thus, in this phase or stage as knowledge about environment and ecology became more widespread, grassroots environmental NGOs gained more influence among the public in China than in former phases or stage. It made their social value more significant than the phase when they existed despite having little influence.

Their financial resources are mainly from foundations in the developed countries. There were few Chinese corporations that provided them with financial resources, except a few that supported events in the first stage or in the first and the second phase. Alashan SEE, established in 2004, was a very unique and important NGO when discussing this topic, because there were 100 corporations in China (including Hong Kong and Taiwan) which financed and supported it. Alashan is the name of a place in Nei Menggu (Inner Mongolia), which is located in the center of a desert. At first, their objective was planting trees in such places facing severe circumstances that could not be solved with the rich capital, provided by plenty of funds. So they changed their main activities, and this was one of the factors associated with the change from this phase to a new one.

10.3.4 The Fourth Phase (2008–): Changing System and Strategy and External Influence

The keywords of the change, the words that characterized the difference around 2008, were system and strategy. Some financial systems were constructed among grassroots environmental NGOs. Domestic and international economic systems[7] were identified by some grassroots environmental NGOs, and they had developed strategies in improving the environment effectively. Though the governmental policy for NGOs increased unstability over time, at least some parts of the cycle between governmental policy and the activities of grassroots environmental NGOs were fixed. The first point of the difference of the terms before 2008 and the present is stated below, while the second and the third points are stated in the next section.

Alashan SEE changed their main activity to finance other grassroots environmental NGOs. Not only their activities but also some domestic financial cooperative relationships among some of the grassroots environmental NGOs were built up rapidly after 2008. For example, a grassroots environmental NGO based in a big city with staff of high-level financial abilities might receive a large sum of funds and donations from an international foundation of developed countries, which could share the funds with many other grassroots environmental NGOs mainly based in small cities or rural areas, which could neither be funded for activities by themselves nor could manage the huge amounts of money generally supplied by international foundations, so this system could suit these organizations.[8]

Commonly, rather than academically, Chinese people categorize young people in China chronologically by the time they were born. For example, if a person was born in the 1980s, he/she is called 80 Hou (meaning post-1980). And they indicate that 80 Hou and 90 Hou are different from the earlier generations, but there is more than one specific thing that characterizes this difference between the generations. One of the explanations related to grassroots environmental NGOs is how to use digital devices for activities. Younger generations are called digital natives or mobile natives, and there is a tendency for the younger generations to be able to use those devices more effectively. I think it is more important that they can also be called environmental natives and NGO natives. Before they were grown up enough to be socialized, they had already absorbed much information about the environment in the media, even with school education, and in some cases the suppliers of the information were grassroots environmental NGOs. For most Chinese people, both the environment and NGOs are new concepts that emerged in the 1990s. For

[7]Here, "economic system" means the whole value chain including supply chain, where pollution sources are generated. Initially NGOs accused each polluter company, but the pollution sources were generated in value (or supply) chains, so they decided to set their target on the whole chain. Detailed information is given in Sect. 10.4.4.

[8]This paragraph is based on an interview with Wen Bo who had responsibility in an International NGO called Pacific Environment at that time in China and owned much information about grassroots environmental NGOs in China, on April 1, 2009.

people who were already an adult at that time, these were alien concepts, but for people who were children at that time and learned about these things from a young age, they accepted them as a normal part of their daily life. These concepts seem very normal and even familiar. To get a job with an NGO may seem unusual for older generations, but not for the young generations.[9]

Some of the younger members of environmental NGOs indicated that their motivation for participating in environmental activities came from Japanese anime movies, especially films directed by Miyazaki Hayao and produced by Studio Ghibli,[10] which may have shown that the Japanese soft power has influenced the young people of China in improving the environment of their country and, as a result, strengthening Japanese environmental security.

From the first to the third, each phase lasted for 5 years. Is it possible that only the fourth is longer than the others? If there was an epoch-making change in 2012 or 2013, the new phase must have already begun. If not, the same phase would have continued. There have surely been some changes, but it is difficult to confirm that they were epoch-making in the present situation because an epoch-making change cannot be recognized immediately. In fact, the year when I recognized that the phase until 2007 and the phase from 2008 were not same, was 2009. If the situation is the same as the current phase, I should have recognized the difference in 2014, which I have not. Of course, this does not constitute any evidence that the phase is the same or has changed.

Some people in Western countries may expect that Chinese grassroots NGOs will make progress toward the democratization of China. But it may be hard for them to realize making a change toward democracy in Chinese society. The most important factor of democracy is bottom-up decision-making, which reflects the willingness of the members of society. But much of the Chinese NGOs' decision-making is not like that; one or two persons provide strong leadership, and the others are just their followers. It is not a simple dictatorship, but it has a top-down tendency, rather than that of a democracy. In fact, Chinese grassroots NGOs' characteristics tend to reflect the character or personality of the leaders. So they may be able to offer an alternative to the Chinese communist party, but there is little likelihood that they will lead to democratization.

---

[9]This paragraph is based on participant observation of Japan and China environmental NGOs' exchange project, especially since 2010.

[10]This sentence is based on participant observation of Japan and China environmental NGOs' exchange project and interviews with Chinese participants of the project, especially younger generations, since 2013.

10.4 Grassroots NGOs and Environmental Pollution Prevention

With the phases described above changed, a new way for grassroots environmental NGOs to prevent environmental pollution appeared. Even as early as the first phase, they recognized environmental pollution as a problem, though they could do nothing about it. In the second phase, the Center for Legal Assistance to Pollution Victims (CLAPV) was established and took actions to provide several kinds of legal assistance and services for environmental pollution victims. CLAPV has taken part in the networks among grassroots environmental NGOs, where their action was welcomed, and some of them followed CLAPV's action. In the third phase, local NGOs like the Guard of Huai River (淮河卫士) in Zhoukou in Henan province and the Green Hanjiang in Xiangfan in Hubei province visited the sites affected by environmental pollution which can cause serious health problems such as the cancer villages, which were broadcasted by the media and the Internet. They also collected social concerns and provided direct support to victims. In the fourth phase, the Institute of Public and Environmental Affairs (IPE) had led the other NGOs through the corporations based on economic and financial relationships, such as supplying chains, so as to force polluters to stop their pollution.

10.4.1 The First Phase: Only Recognition

Before 1998, at least for FON, it was recognized that environmental pollution must be prevented. However, they were unsure how to nip the problem in the bud under the strict Chinese political system. As a result, they became helpless to change or influence the way in addressing or preventing environmental pollution.[11]

10.4.2 The Second Phase: The Beginning of Activities to Prevent Environmental Pollution with Legal Assistance

The Center for Legal Assistance to Pollution Victims (CLAPV) obtained permission for its establishment in 1998 and started its activities in the following year. Their core action is to provide consultations for environmental pollution victims by means of telephone, letters, and visit without charging a fee. They choose severe and important cases they have already provided with consultations and then offer

[11]This sentence is based on interviews with the late Mr. Liang Congjie who was the leader of FON at that time in August 1994 and January 1999.

several types of legal assistance such as representing the victims in suits. Cases which have been provided with legal assistance were approximately one percent until 2010. It has increased since the year it reformed, but hasn't doubled what it was before since most consultations are not about severe pollution cases but are common disputes among neighbors, and CLAPV is limited in handling these cases.

CLAPV also conducts legislative proposals and training for lawyers, judges, and administrators. It aims to prevent environmental pollution by reforming the whole legal system, including the law itself and its operation system through these activities. In recent years, CLAPV has even provided NGO staff with training.

CLAPV started exchange programs with other grassroots NGOs, influencing each other interactively. Besides, the media showed interest in CLAPV's activities and sometimes broadcast them, which also increased CLAPV's influence. Their formal exchange with the Japanese organizations began in 2000 as detailed in the next section, which coincide with the two activities as mentioned above.

Influenced by CLAPV, other organizations might have begun their activities in the following years of the second phase. For example, circa 2003, FON and Greener Beijing took action in supporting environmental pollution victims of Inner Mongolia with the advice of CLAPV. At that time, some promotional videos made by FON began to show concern with supporting actions for environmental pollution victims (Aikawa 2005).

10.4.3 The Third Phase: Revealing Local Pollution Problems and Supporting Victims

In the third phase, local NGOs of rural areas became the new and main organization to prevent environmental pollution, while in the second phase, NGOs in one of the biggest urban areas, Beijing, the capital city, continued their activities and supported their local colleagues in the third phase.

Mr. Huo Daishan, a photojournalist and the establisher of the Guard of Huai River, revealed that there are many "cancer villages" (*Zhongliu Cun*, *Aizhong Cun*, or *Aizheng Cun*) in Zhoukou City in Henan province which constitutes a part of the Huihe River (Huai River) valley. Many residents in this area suffer from cancer and other health problems since the late 1990s, which could be caused by industrial water pollution. His appeal was reported broadly not only within China but also abroad in 2004 – 10th anniversary year of the fully fledged prevention of Huaihe River water pollution. The entire Chinese society including the government was shocked by the revealed situation. All of these helped him and his organization obtain a lot of support from others, enabling them to conduct more research. And thus they found the scope of "the cancer villages" was wider than they had known before (Aikawa 2005:25–27, 2006:165).

Green Hanjiang in Xiangfan in Hubei province found and exposed that there were some villages in a tributary valley of the Hanjiang River (which is also a tributary of the Yangtze River) having many cancer patients when they cooperated with a governmental campaign to protect rivers. These villages were called the "cancer villages." Their disclosure also attracted the attention of NGO networks and mass media greatly.

Chinese society was deeply shocked by the current condition of the Huaihe River valley and Hanjiang River valley, and a support movement from different levels of society broke out.

In the meantime, many accidents and incidents concerning with environmental pollution occurred in succession produced a significant impact on the Chinese society. These include the Tuo River Pollution Accident (spring of 2004), at least four outbreaks of protests against pollution, antipollution mobs in Zhejiang province (spring of 2005), the Songhua River Pollution Accident (autumn of 2005), the Bei River Pollution Accident (winter of 2005), the multiple episodes of explosive breeding of blue-green algae by cyanobacteria at Taihu Lake and many other areas (summer of 2007), among others. These accidents and incidents let several grassroots environmental NGOs which had already attracted by the environmental pollution become more interested in these areas, especially about water pollution. In 2006, Japanese, Chinese, and South Korean NGOs and citizens held a conference[12] with the name of "water and health," but the real topic was water pollution.

By 2007, the final year of this phase, the Guard of Huai River had succeeded in building a cooperative relationship with the corporation that they had criticized as the most suspicious polluter. The corporation's name also indicates the model of the relationship, which is called the "*Lianhua* model." This is not only a model for their organization of the activities in their area but also for general cooperative relationships between NGOs and corporations, especially for grassroots NGOs.

In the last term of the third phase, there were several new movements related to the new phase. In 2006, the Institute of Public and Environmental Affairs (IPE) was established. It collected official information about water pollution and published it on the Internet. It was called a "water pollution map," but it is actually a blacklist of water polluters. They expected consumers could use this list primarily. In order to gather these consumers together and inspire movements and campaigns, they established an objective network named as the Green Choice Alliance (GCA) with other grassroots environmental NGOs.

[12]The 3rd East Asia Environmental Citizen Conference.

10.4.4 The Fourth Phase: Focusing on Economic and Financial Relationships

Corporations could be sponsors for grassroots environmental NGOs from the beginning, even though that was not true for most cases. Since CLAPV began their activities, the other possibility for corporations had become the object of criticism or as an enemy of grassroots environmental NGOs for a decade. It was a very simple and extreme relationship.

Before 2008, the main target of activities to prevent environmental pollution by grassroots NGOs had been individual polluting corporations, while an important exemption being the whole legal system that is targeted by CLAPV. After that year, grassroots environmental NGOs began to adopt more strategic approaches.

Influenced by policy suggested by the grassroots environmental NGOs network under the leadership of CLAPV and IPE in 2007, the Chinese government established a series of institutions for information disclosure in 2008. It enabled every grassroots environmental NGO, including those that had not participated in making the suggestions, to obtain much more information than before. Because of the new institutions, they accused the polluting corporations directly and indirectly, even some of them were from abroad. In some cases, grassroots environmental NGOs succeed in having dialogues with the polluting corporations or related corporations. In addition to the "water pollution map," the IPE also opened up an "air pollution map" in the Internet, and it is in fact also a blacklist. Mr. Ma Jun, the leader of IPE, said that some corporations did not want their name to be exposed on the blacklists. Therefore, they contacted IPE. And this was a chance to start a dialogue (Aikawa 2008c:225). Of course, their objective has not been merely the dialogue, but improving the situation through frequent communications, aiming to solve the pollution issues eventually.

At the meantime, since the disclosure had provided grassroots environmental NGOs with too much information, they were forced to make it more effectively. Even before 2008, the number of companies on the blacklist, which was made by the IPE and used by the GCA, had increased to thousands, and mobile devices at that time were not as popular as the time after 2010 in China, so it was hard for general consumers to purchase goods in a shop according to the blacklist. So the members of the IPE and GCA began to focus on supply chains. They set retailers and final manufacturers, especially regarding famous corporations as the new target. They must choose suppliers, but if one is a polluter they must not buy from it. This is almost like a duty for corporations, especially when the corporation professes itself as a company with "social responsibility."

The movement expanded from some special cases of polluters to an entire industry, from the entire industry to several industries, and from supply chain to value chain. The meaning of the movement growing from supply chain to value chain is that not only buyers but also investors and financiers became the targets of the movement.[13]

[13]This paragraph is based on the interviews with persons who are related to IPE and GCA in Beijing in August and September 2014.

It is difficult to estimate how much of these activities and movements could act in preventing environmental pollution in China. Every activity discussed herein has certain strong impacts on the corporations and society. Of course, it cannot be assessed as perfect or enough because environmental pollution is still in a difficult situation to handle at this stage.

10.5 Relationship with Japanese Civil Society: Its Influence on and Exchange and Cooperation with Chinese Grassroots Environmental NGOs

Japan has experienced some serious environmental pollution in the world, especially among developed countries. In China, since the 1970s, people from different social levels including the top leaders and the ordinary people, have shown interest in the environmental pollution Japan faced. Grassroots environmental NGOs were no exception, and even before the beginning of their activities, some leaders and members had already paid much attention to it, and they started to initiate some exchange programs and cooperate with the Japanese civil society.

Japanese civil society, especially the anti-*Kogai* (environmental pollution) movement, paved the way for the enactment of effective environmental policies in Japan in the 1960s and 1970s. They also paid attention to the environmental situation in China. All of these have driven their efforts to develop exchange programs with China and thus providing an opportunity to start cooperating with the Chinese civil society as early as 1973 – just one year after the normalization of Sino-Japanese relations. Despite their success, they had to wait for an opportunity for more than 20 years because there was almost no opening for them to penetrate into the Chinese civil society in the environmental area. Moreover, even if there was an opportunity, it was hard for them to access before the establishment of FON.

10.5.1 Brief History of Exchange and Cooperation

Leaders of the Chinese government were interested in the experience of Japan and ordered their staff to study and learn about it around 1970. In the 1970s some Japanese specialists on environmental pollution such as Ui Jun and Dr. Harada Masazumi, the famous researchers of the Minamata disease, visited China and communicated with people from different social levels. The leaders of grassroots environmental NGOs were also interested in Japan's environmental experience. The leader of the Guard of Huai River, Huo Daishan, learned about *Kogai* in Japan in the early 1980s, especially the case concerning the Minamata disease. Subsequently, he recognized the situation in the Huaihe River valley was also *Gonghai* (*Kogai*). The leader of CLAPV, Wang Canfa, at the first meeting with two young

Japanese researchers (including me) in 1998, suggested to hold workshops with the Japanese researchers, scholars, and lawyers. And this was merely the time CLAPV had been permitted to form, without performing any activities. His suggestion was realized three years later.

The workshop was an important event in the second phase, and it was the first time that Chinese pollution victims spoke out publicly and made speeches at an international meeting, where some Chinese officials were also in attendance. Even now this situation rarely occurs. Victims also guided field trips and showed the actual circumstance they faced with regard to environmental pollution. And these gave Japanese participants, especially those who had visited China before, including specialists such as Ui Jun and Dr. Harada Masazumi, an impression that the situation in China was improving. The Japanese counterpart was the Japan Environmental Council (JEC), a unique NGO that was organized by various specialists including lawyers, members of citizen movements, environmental pollution victims, and supporters. However, their network did not confine to these types of people. It activated and accelerated interactions and cooperation in the field of environmental pollution between Japanese and Chinese civil societies.

In the third phase, the following-up workshops which included the participants from South Korea, were held for three times, in 2004, 2005, and 2007. Based on the workshops where CLAPV broadened its interaction with Japan, the workshops provided the impetus and lessons for China to partially amend its law in order to protect the rights of the environmental victims. There were site-visit and environmental conference organized by Japan and China other than CLAPV. These included JEC members, including Dr. Harada Masazumi, visiting the Huaihe River valley in 2005 and the 3rd East Asia Environmental Citizen Conference in Xi'an held in 2006. But in the last two years of this phase and the following years, some aspects of the situation had also become worse. For example, sometimes Japanese visitors were forbidden to access the polluted sites, where they had been welcomed and guided by the locals earlier. Above all, perhaps the third phase was the peak of these interactions and cooperation.

Through the third phase, the people who learned the most from these interactions were almost always on the Chinese side. But in the fourth phase, in some cases, the Japanese became the one who had learned something from the Chinese as in the field of green supply chain and green finance. Regarding the workshop between CLAPV and JEC, instead of holding a workshop, there was a book published in two languages, in Japanese (Japan Federation of Bar Associations, Environmental Protection Committee ed. 2010) and Chinese (日本律师协会 主编2011), introducing the roles played by lawyers in the legal conflict of environmental pollution suits in Japan. The book was edited because it was requested by CLAPV, but there hadn't been any book like that in Japan, so the publication was valuable in Japan, as well as in China.

10.5.2 Consideration About "Traps"

Japanese NGOs are different from other developed countries' NGOs. Their powers are weak, and they always face difficulties in fund-raising. Besides, their members prefer to pay attention to the hot issues of environmental problems rather than to pay the cost of organization or donations for public benefit. Because many Japanese do not customarily give donations, with only a few exceptions like for natural disasters, the Japanese NGOs are financially poor. Many of their members want to act at the site, and they want to pay for themselves, but they do not want to pay too much for the organizations or staffs. Even Greenpeace, which owns a principle that they must initiate fund-raising in developed countries, cannot raise enough funds in Japan to support itself and needs support from overseas.

In fact, Japanese NGOs have been trapped in almost the same situation described above for about two decades. As Chinese NGOs develop, they are likely to be able to catch up easily, and partly they have already been aligned. But what will happen in the future? It is hard to predict, as there will be some "traps" in China, too. For example, they say Chinese economics is caught in the "middle income trap." And if that is true, what about the NGO sector? Questions like this will be analyzed in the coming years. However, as a predecessor, Japanese NGOs have to seek ways to avoid them and continue to grow. This will help to improve the situation of solving environmental problems not only in Japan but also in China.

10.6 Conclusion and Update

It has been over 20 years since the first grassroots environmental NGO was established in China. With the emergence of more and more grassroots environmental NGOs, their actions, active areas, members, types of organizations, specialties, and established locations have become more diversified since then. One of the additional fields of their activities was the prevention of environmental pollution. They have also developed networks among themselves. Some of the environmental problems were revealed by grassroots environmental NGOs, so their social influence has strengthened since 2003. Their projects focus on exchanges and cooperation with environmental NGOs, environmental pollution victims, and supporters in Japan that are also expanding. In order to deal with environmental problems efficiently in China, civil societies in Japan and China are expected to be achieved.

After finishing this chapter on September 2015, I visited China and met with several members of grassroots environmental NGOs. During this visit, I noticed that there were two points needed to be added.

The first point is that the independence of grassroots NGOs may be endangered with the changes of policy. In the latter half of 2014, the communist party began to require every NGO owning a communist party branch within the organization. This means that every NGO must, in many cases, accept one or more members who are also communist party members. But the truly independent organizations are organizations which can make their own rules. Some said that the same policy had existed before, so this was only a reminder. But others said that this was almost a command and was absolutely not the same as it was before. It is hard to say which is right, but we must pay attention to the future situation whether this policy will bring some changes or not.

The second point is that the redevelopment of pollution-damaged areas has become a social issue in China, and some grassroots environmental NGOs have begun to initiate activities in this area. This point will be left to another opportunity to be fully detailed.

References

Aikawa, Yasushi [2005] "Environmental Damages, Disputes and NGOs' Activities in China", Annual Bulletin of Research for Social Sciences, No. 35, pp. 17–30 (in Japanese).

Aikawa, Yasushi [2006] "Environmental Pollution in China Getting "Explosively" Worse", Bulletin of Tottori University of Environmental Studies, Vol. 4 Feb. 2006, pp. 159–178 (in Japanese).

Aikawa, Yasushi [2008a]'Environmental NGOs in China', Kitagawa, Hideki ed. "Managing China's Ecological Crisis: the Current State of Environmental Law and Policy", Houritsubunkasya, pp. 315–343 (in Japanese).

Aikawa, Yasushi [2008b] "Environmental NGOs in China and the Subject of Japan-China Environmental Cooperation", Environmental Research Quarterly, No. 150 Aug. 2008, pp. 45–52 (in Japanese).

Aikawa, Yasushi [2008c] "Environmental Pollution in China", Softbank Creative (in Japanese) .

Aikawa, Yasushi [2012] 'NGOs' Activities for Prevention of Environmental Pollution in China', Kitagawa, Hideki ed. "Environmental Law, policy and governance in China" Koyoshobo, pp. 127–141 (in Japanese).

Bao, Maohong [2009], Translation supervised by Kitagawa, Hideki, "Environmental Governance in China and Environmental Cooperation in Northeast Asia" Harupress (in Japanese).

Economy, Elizabeth C. [2004], The River Runs Black: The Environmental Challenge to China's Future, Cornell University Press.

Japan Federation of Bar Associations, Environmental Protection Committee ed. [2010] "Environmental Suites and Challenge of attorneys", Houritsubunkasya (in Japanese).

日本律师协会 主编 [2011]、王灿发 监修、皇甫景山 翻译《日本环境诉讼典型案例与评析》中国政法大学出版社 (Chinese version of Japan Federation of Bar Associations, Environmental Protection Committee 2010).

Turner, Jenifer L., Lv, Zhi "China's Environmental NGOs", Center for Environment and Development, Chinese Academy of Social Sciences, China Environment and Development Review, Vol. 3, Social Sciences Academic Press (China), pp. 349–369 (in Chinese) =特纳,珍妮弗L. 、吕植 [2007]「中国环境NGO的作用」中国社会科学院环境与发展研究中心『中国环境与发展评论』社会科学文献出版社、349–369页.

Index

H. Kitagawa (ed.), *Environmental Policy and Governance in China*,
DOI 10.1007/978-4-431-56490-4

MIX
Papier | Fördert gute Waldnutzung
FSC® C083411

Kolibri 360 GmbH
Ferdinand-Jühlke-Straße 7
99095 Erfurt, Deutschland
produktsicherheit@kolibri360.de